ROAD ASSET MANAGEMENT SYSTEMS AND PERFORMANCE-BASED ROAD MAINTENANCE CONTRACTS IN THE CAREC REGION

NOVEMBER 2021

CAREC
Central Asia Regional Economic Cooperation Program

ADB

Notes:
In this publication, "$" refers to United States dollar, "CNY" to Chinese yuan, "GEL" to Georgian lari, "MNT" to Mongolian togrog, "Som" to the Kyrgyz Republic som, and "T" to Kazakhstan tenge.
ADB recognizes "China" as the People's Republic of China.

On the cover: Road maintenance works at a portion of a highway at Zhambyl Region in Kazakhstan under the CAREC Transport Corridor 1 (Zhambyl Oblast Section) Investment Program and in Garm district, Tajikistan under the Dushanbe–Kyrgyz Border Road Rehabilitation Project (photos by Sean Crowley and Nozim Kalandarov for ADB).

Contents

Table and Figures

Table

Figures

Acknowledgments

This publication on road asset management systems (RAMS) and performance-based road maintenance contracts (PBCs) in the Central Asia Regional Economic Cooperation (CAREC) region was developed as part of the support of the CAREC Program toward the establishment of competitive transport corridors; facilitation of movement of people and goods; and provision of sustainable, safe, and user-friendly transport and trade networks. This publication was prepared in the context of CAREC's Transport Strategy 2030 and aims to improve the management and maintenance of CAREC road corridors through the sharing of best practices that exist within CAREC countries regarding RAMS and PBCs.

Serge Cartier van Dissel, road management consultant for the Asian Development Bank (ADB), prepared this publication. Thanks goes to the different people who contributed information on the experiences with RAMS and PBCs in various CAREC countries. Special thanks go to Thomas Hertz, Ritu Mishra, and Pilar Paz Sahilan (consultant) of ADB, for their support in collecting information, and for their contributions to and detailed review of the various versions of this document. Last but certainly not the least, the contributions received from the staff of the different road agencies and ADB resident missions in CAREC countries must be recognized—those who provided detailed information on the experiences with RAMS and PBCs in their respective countries; who participated in discussions on these topics; and who assisted in the review of this document, correcting and complementing the information where necessary.

Abbreviations

ADB	Asian Development Bank
AFG	Afghanistan
AMP	Annual Maintenance Plan
AZE	Azerbaijan
BCI	bridge condition index
CAREC	Central Asia Regional Economic Cooperation
cm	centimeter
COR	Committee of Roads
CPMS	China Pavement Management System
DDH	Департамент Дорожного Хозяйства—Road Maintenance Department
DEU	дорожно-эксплуатационное учреждение—road maintenance unit
DOR	Department of Roads
dTIMS	Deighton Total Infrastructure Management System
EBRD	European Bank for Reconstruction and Development
FWD	falling weight deflectometer
GDP	gross domestic product
GEO	Georgia
GIS	geographic information system
GPR	ground penetrating radar
GPS	global positioning system
GUSAD	Главное управление строительства автомобильных дорог—State Institute for Road Maintenance
HDM4	Highway Design and Maintenance, version 4
IRI	International Roughness Index
JICA	Japan International Cooperation Agency
KAC	Keep Afghans Connected
KAZ	Kazakhstan
KGZ	Kyrgyz Republic
km	kilometer
LCMS	laser crack measurement system
m^2	square meter
MAC	Management Agency Contract
MIID	Ministry for Industry and Infrastructure Development
MON	Mongolia
MOT	Ministry of Transport
MOTC	Ministry of Transport and Communications
MOTR	Ministry of Transport and Roads
MPW	Ministry of Public Works
MQI	maintenance quality index
MRTD	Ministry of Roads and Transport Development
MS	Microsoft
NCQRA	National Center for the Quality of Road Assets
NHA	National Highway Authority
NPS	Network Performance Score

NRAP	National Rural Access Program
OPRC	output- and performance-based road contract
PAK	Pakistan
PASER	Pavement Surface Evaluation and Rating
PBC	performance-based road maintenance contract
PCI	pavement condition index
PIC	Производственно-инновационный центр—Production Innovation Centre
PLUAD	Производственно-Линейное Управление Автомобильных Дорог— Production and Linear Management of Roads
PMS	pavement management system
PPP	public–private partnership
PRC	People's Republic of China
PSSI	pavement structure strength index
PWI	pavement surface wearing index
RAIP	Road Access Improvement Project
RAMD	Road Asset Management Division
RAMP	Road Asset Management Project
RAMS	road asset management system
RDI	rutting depth index
ROMDAS	Road Measurement Data Acquisition System
RONET	Road Network Evaluation Tool
RQI	riding quality index
RRF	Republican Road Fund
RTDC	Road and Transport Development Center
RTSD	Road Transport Service Department
SLA	service-level agreement
SQL	Structured Query Language
SRI	skid resistance index
SRN	strategic road network
TAJ	Tajikistan
TKM	Turkmenistan
UAD	Управления Автомобильных Дорог—Management of Roads
USAID	United States Agency for International Development
UZB	Uzbekistan
VIMS	Vehicle Intelligent Monitoring System
YHAB	Yunnan Highway Administration Bureau

Introduction

This document provides an overview of the development status of road asset management systems (RAMS) and performance-based road maintenance contracts (PBCs) in the 11 member countries of the Central Asia Regional Economic Cooperation (CAREC) Program,[1] and identifies the way forward and next steps to be taken in the development of RAMS and PBCs. This document builds on earlier efforts by the Asian Development Bank (ADB) in support of introducing RAMS and PBCs in the CAREC region. This includes the *Compendium of Best Practices in Road Asset Management and the Guide to Performance-Based Road Maintenance Contracts* published by ADB in 2018. In parallel with the current document, *CAREC Region Road Asset Management Maturity Assessment* was prepared by the CAREC Institute, which covers several aspects of RAMS and PBCs and complements this document.

For each CAREC member country, this document provides an overview of the road network, the institutional responsibilities for its management, the involvement of government units and private sector contractors in the implementation of road works, and the levels and sources of funding for the road sector and especially for road maintenance and repairs. The terminology used for the different road treatments is also explained to facilitate the comparison between different countries. This provides the overall framework within which RAMS and PBC development is assessed.

The status of RAMS development in each CAREC member country is subsequently described, listing the different actions that have been undertaken and are ongoing, often with development partner support. The definition of a RAMS applied in this document is: "any system that is used to collect, manage and analyze road data for road planning and programming purposes."[2] RAMS assessment in this document looks at the three main elements: (i) data collection, (ii) database management, and (iii) data analysis. The degree of RAMS integration into the wider framework is also reviewed, as this tends to be a weak area in RAMS development, leaving the RAMS unconnected to overall road management. This looks especially at the institutional framework for RAMS operation, the use of the RAMS in existing planning and budgeting procedures, its impact on financing levels and budget allocations for road maintenance and repair, and the existing domestic capacity to carry out planned road maintenance and repair works.

This document also describes the status of PBCs in each CAREC member country, describing the different PBC pilots that have been implemented or attempted. The definition of a PBC applied in this document is: "any road contract or agreement that involves payments on a performance basis (against compliance with specific road condition targets)."[3] Particular attention is given to the system of performance standards that form the basis of any PBC and define the road condition to be achieved. PBC assessment also looks at the system of payment deductions that is applied in case the performance standards are not complied with. The use of response times is also given consideration, as this has led to problems in the implementation of PBCs around the world.

---

[1] CAREC member countries include Afghanistan, Azerbaijan, the People's Republic of China, Georgia, Kazakhstan, the Kyrgyz Republic, Mongolia, Pakistan, Tajikistan, Turkmenistan, and Uzbekistan.

[2] ADB. 2018. *Compendium of Best Practices in Road Asset Management.* https://www.adb.org/sites/default/files/publication/396126/best-practices-road-asset-management-carec.pdf.

[3] As opposed to payments on an output basis (volume of work completed) or input basis (amount of resources used).

Apart from providing a description of the current status of RAMS and PBC development, this document also provides information on the way forward in CAREC member countries. For each country, the next steps to be taken to further develop RAMS and PBCs are described, based on current status and immediate issues being faced.

Although some elements are specific to a particular country and its overall framework, there are elements common for different countries, especially those with similar levels of RAMS and PBC development. In the final chapter, particular attention is therefore given to these common elements that strongly influence the success of RAMS and PBC development, building on the lessons learned from previous assessments carried out by ADB and other organizations. This starts with an overview of RAMS and PBC development in the CAREC region, assessing the overall progress made in each country, and comparing the levels of development. Although there are considerable differences, these reflect how far along each country is in the typical three phases of piloting, replication, and integration, and is to a large extent dependent on when the development of RAMS and PBC was started.

Afghanistan

The assessment presented in this chapter reflects the situation at the beginning of 2021 and does not take into account the recent changes.

Road network. Afghanistan has a strategic road network (SRN) of 19,327 kilometers (km), including 6,854 km of regional and national highways, 1,394 km of provincial roads, and 11,079 km of district roads. According to the Ministry of Public Works, considerable additional length is under construction, in the pipeline, or planned. In terms of road activities, distinction is made between routine maintenance, periodic maintenance, and rehabilitation and (re)construction.

Institutional framework. The management of the SRN used to be the responsibility of the Ministry of Public Works (MPW). In December 2018, the MPW was converted into the National Road Authority under the newly established Ministry of Transport (MOT). The National Road Authority did not function as the semiautonomous entity it was foreseen to be and has since been converted back to being the MPW. Most roadworks used to be carried out by in-house units at the provincial level, but these units now focus only on emergency and winter maintenance. Routine and periodic maintenance of paved roads are contracted out to contractors, while routine maintenance of unpaved roads is largely carried out by microenterprises and community development councils. Due to a lack of private sector capacity, an in-house *Qway-e-kar*[4] unit was established to carry out routine and periodic maintenance and complement the work carried out by the private sector. A Road Fund Unit was recently created under the Ministry of Finance, which is responsible for managing revenues from road user charges and their allocation to the road sector.

Financing. According to the Road Sector Strategy 2019–2023 of the government, 42% of the SRN requires urgent periodic maintenance and a further 10% requires rehabilitation or reconstruction. Between 2011 and 2014, the Government of Afghanistan spent on average $23 million per year on road maintenance, accounting for less than 5% of total road subsector expenditure. The road sector strategy estimates that two-thirds of the paved SRN roads will require periodic maintenance in 2019–2023, with a total budget requirement of $143 million. Routine maintenance involves an additional budget requirement of $29 million, while for emergency maintenance the cost is estimated to amount to $50 million for the same period. The unpaved SRN roads are expected to require an additional $55 million, mainly for routine maintenance by microenterprises and community development councils to keep the roads passable. This is equivalent to a maintenance budget of just over $55 million per year, double the current budget allocation. The Transport Master Plan Update 2017–2036[5] of the Asian Development Bank (ADB) estimated the maintenance requirement for the SRN to be much higher, with an annual budget requirement of approximately $250 million, including $70 million for routine maintenance and $180 million for periodic maintenance. This is similar to the budget estimated by the Short-Term Plan for Interim Funding of Road Maintenance, funded by the United States Agency for International Development (USAID). The ADB-funded Road Asset Management Project provided a middle ground, estimating in 2019 that an annual budget of $100 million was required for the maintenance of SRN roads. Although maintenance financing is a limiting factor, a lack of sufficient management capacity to organize the implementation of works means that a portion of the budget is left unused each year. The government aims to significantly increase its capacity over the next few years so that this no longer forms a constraint.

---

4   *Qway-e-kar* is an implementing arm of the Ministry of Public Works that focuses on periodic and routine maintenance.
5   ADB. 2017. *Afghanistan Transport Master Plan Update 2017–2036*. TA 8687-AFG. Manila. https://www.adb.org/documents/afg-transport-plan-update-2017-2036.

Road user charges. Afghanistan introduced a fuel tax with revenue earmarked for road maintenance and allocated to a special account under the MPW along with revenue from tolls and traffic fines. In practice, however, there has been uncertainty whether the annual revenue allocations to MPW match the earmarked revenues. In the past, ADB and the USAID supported the establishment of a semiautonomous Road Fund, preparing the draft legislation for its establishment, but this has yet to be created. A Road Fund Unit was recently created under the Ministry of Finance as an alternative to a semiautonomous Road Fund. The Road Fund Unit receives earmarked revenue from road user charges for road maintenance and development. The road user charges to be included still need to be defined but will likely include the existing fuel tax as well as an annual vehicle road tax.

Road Asset Management System

Development. RAMS development in Afghanistan has been supported by a variety of projects and programs funded by different development partners. This includes the USAID-funded Road Operation and Maintenance and Capacity Building Program (Task Order 14, 2007–2011), the Japan International Cooperation Agency (JICA)-funded Establishment of a Road Maintenance and Management System (2008–2012), the Swedish International Development Agency-funded Rural Access Improvement Program (RAIP, 2007–2015), the Department for International Development of the United Kingdom-funded Keep Afghans Connected project (KAC, 2013–2014), the USAID-funded Road Sector Sustainability Project (2014–2016), and the World Bank-supported National Rural Access Program (NRAP, 2012–2020). Up until the end of 2020, RAMS development was being supported under the ADB-funded Road Asset Management Project (RAMP). With support from the RAMP, a Road Asset Management Directorate was created under MPW. A total of eight zonal units were also introduced, each covering several provinces and including specific data collection units. These are supported in the data collection by existing provincial offices.

Data collection equipment. Different approaches have been applied to data collection over the years. Although equipment-based measurements have been recommended, difficulties with access and security in Afghanistan have led to the application of visual assessments. The most successful of these was the Pavement Surface Evaluation and Rating (PASER) applied under the KAC project to evaluate road conditions using a rating from 1 (severe distress) to 10 (no distress). The condition is defined for each rating, and each rating is linked to the required type of treatment (routine maintenance, seal coat, overlay, pavement rehabilitation, or reconstruction). This approach has been approved by the government and was applied under the RAMP. Some basic equipment have been used for data collection, such as global positioning system (GPS)-enabled cameras in the KAC project and use of the smartphone-based RoadRoid application in the RAIP. Due to the poor security situation, vehicle-based survey equipment has not yet been used, although the procurement of survey vehicles with basic equipment is being considered by MPW. This would involve various vehicles in different cities spread over the country to address the difficulties of traveling from one city to another due to security risks.

Data collection. Between 2007 and 2011, JICA collected data for 300 km of roads in Kabul and Bamyan provinces. In 2010, geographic information system (GIS) mapping was carried out for the entire SRN based on high-resolution aerial mapping as part of the Afghanistan Information Data Center with USAID support. This resulted in data on road centerlines (all SRN roads) and large structures (regional roads only). However, this data has not been updated since 2014 and presents discrepancies with current satellite imagery. In 2015, data were also collected in four provinces under the RAIP, including inventory data, general condition data, roughness data, and traffic counts. The KAC project collected data in 2016 using tablets and cameras with GPS functionality, resulting in data for road centerlines (SRN roads) and visual condition assessments (national roads). Under the NRAP, road inventory data were collected for a large portion of the road network in 2017. Although initially focused on tertiary roads, this was later expanded to include secondary and strategic roads. Single-day traffic counts were also conducted in several roads. However, due to a system crash, the data are no longer accessible. Data collection between the different projects was not coordinated and was not always shared. As a result of the haphazard approach, data for the SRN is incomplete and most of it is outdated or not in line with geospatial data standards.

Under the RAMP, over 30,000 km of roads were mapped using aerial images and GPS equipment, as well as GIS data from earlier projects. Due to issues with capacity and security, a decision was taken to apply the PASER ratings for pavement condition surveys rather than road survey equipment, focusing on condition assessments of paved SRN roads. So far, PASER data has been collected for 3,000 km of SRN roads, and the MPW is considering to extend this to include all paved national and provincial roads. Inventory and condition data were also collected for bridges through visual surveys using forms adjusted from NRAP. Condition data were collected regarding the extent and severity of defects for 21 different bridge elements, which were used to calculate the bridge condition index (BCI). So far, data have been collected for 500 structures and the MPW is considering to extend this to include approximately 1,000 structures in national and provincial roads. Data and photographs for pavement and bridge condition were collected using GPS-enabled tablets. Traffic counts were conducted using an adjusted version of the NRAP forms. The RAMP prepared manuals and handbooks for data collection by MPW and provided training to MPW staff. Data collection was initially carried out by consultants under the RAMP, but in the future will need to be carried out under the direction of MPW and implemented by the zonal and provincial offices.

Database. The involvement of various projects and programs has led to various systems being developed, containing different and often conflicting data. The USAID provided different ministries with the ArcGIS software and related data when they handed over the Afghanistan Information Data Center. The KAC project entered the data into a GeoServer. Data collected under the RAIP was stored in Microsoft (MS) Excel and a GIS database. As part of the inventory data collection under the NRAP, an open-source PostgreSQL database was developed, but is no longer functional. MPW staff have recently started using GIS to visualize project locations. However, none of the systems developed allow for a proper integration of database and GIS mapping. The RAMP proposed to use ESRI Enterprise Web GIS complemented by an MS Structure Query Language (SQL) Server database as the basis for a pavement management system (PMS), since this commercial software is continuously updated and properly supported. The MPW is considering procuring a separate off-the-shelf bridge management system (BMS) software. A web-based interface is currently being developed to link to the PMS and BMS. The RAMS will be managed by the Road Asset Management Directorate under the MPW, with remote access from the zonal and provincial offices. The RAMP found that the network capacity of the MPW was adequate, but that the internet connection in the provinces was poor and would likely hamper proper access to the centralized database.

Data analysis. The USAID-funded Task Order 14 made use of Highway Design and Maintenance, version 4 (HDM4) for preparing multi-annual plans in support of its maintenance program. HDM4 training was also provided to staff of the Road Management Unit set up under the MPW, but this unit was dissolved after the project ended. Under the Road Sector Sustainability Project, the Road Network Evaluation Tool (RONET) was introduced as being easier to use than HDM4. The later NRAP again made use of HDM4, although its use was limited to project consultants and little capacity building was carried out. The RAMP reviewed the option of using HDM4 but determined that existing staff capacities in the MPW and available data were not adequate for it to be used. RAMP also reviewed the option of using Roadsoft, the PASER-related software, but found this to use an outdated GIS platform. RAMP instead suggested the use of basic algorithms in the medium term, with the aim of moving toward the use of HDM4 within the next 5 years. The basic algorithms applied by RAMP give priority to routine maintenance of roads in good condition (PASER ratings 8–10), followed by periodic maintenance of roads in fair to poor condition (PASER ratings 4–7)[6] whereby priority is given to roads with higher PASER ratings to avoid further deterioration. Average costs per kilometer were determined for the treatments linked to each PASER rating, differentiating between different classes of roads. Within each PASER rating, roads are prioritized based on the estimated cost divided by the traffic volume and the length of the road (as a proxy for the cost–benefit per road user). The impact of traffic volumes on road deterioration and road user costs is not yet taken into account. For the prioritization of bridge maintenance, the BCI is used to calculate the priority condition index. Due to the lack of historical treatment data as a basis for predicting future deterioration, planning is currently limited to annual plans. These plans also take account of the security situation, postponing works in areas with serious security concerns. Data analysis will in future be carried out by the Road Asset Management Directorate under the MPW.

Performance-Based Road Maintenance Contracts

Pilots. Various PBC pilots have been carried out in Afghanistan. In 2006, the European Union (EU) awarded a PBC for 3 years of routine and winter maintenance of the 142 km Kabul–Jalal Abad road following its rehabilitation. The contract was awarded to a domestic company at a cost of approximately $2,800/km/year. Between 2007 and 2012, the USAID through its Road Operation and Maintenance and Capacity Building Program (Task Order 14) introduced performance-based contracting for routine and emergency maintenance of 1,626 km of regional and national highways by Afghan contractors. The works were contracted out in five lots to three contractors for an implementation period of 26 months, with an average cost of $6,300/km/year. The costs were higher than the EU pilot but involved roads that were not necessarily recently rehabilitated. An evaluation recommended the use of smaller contract sizes (100–150 km) to improve effectiveness and reduce risks for contractors. Even those road lengths are considerable when first introducing PBCs, and small contracts (or lots) should be considered initially and gradually scaled up, while still ensuring that contract sizes are sufficiently large to be of interest to contractors. The contracts were managed by the Road Management Unit set up under the MPW and staffed by the program. With the end of the program in 2012, the PBCs and the Road Management Unit also ended. More recently, the NRAP introduced 237 road maintenance microenterprises consisting of local community members who were paid on a performance basis for the routine maintenance of 3,600 km of rural roads, generating 3.4 million person-days of employment up to 2020. Under the Road Sector Strategy 2019–2023, this approach was planned to be extended to unpaved SRN roads.

Performance standards. An evaluation by the USAID of the Road Operation and Maintenance and Capacity Building Program found that the performance standards used in PBCs were unclear. This resulted in difficulties for the contractors to bid for the contracts, but also resulted in varying results from the monthly inspections. The security issues in Afghanistan formed a major impediment, and to a large degree the inspections depended on aerial and other photographic evidence.

---

[6] Roads with lower PASER ratings require rehabilitation or reconstruction and are not included in maintenance planning of roads and highways.

Way Forward

Road asset management system. RAMS development in Afghanistan has taken account of the lessons learned from previous experiences, the existing capacities of MPW staff, and the impact of security issues, focusing on the development of a system that could be operated by the MPW and that allowed evidence-based planning to be carried out. The main challenge is the institutionalization of the approach, ensuring that the MPW will replicate the data collection in future years and will use the RAMS as the basis for their maintenance planning. A secondary challenge is to introduce simple survey equipment for objective measurement of road conditions, ranging from simple smartphone applications to basic vehicle-based bump integrators and video cameras with GPS linkages (such equipment can be installed inside the vehicle, and thus has lower security risks). This will facilitate the collection of data and allow further analysis to be carried out, using basic algorithms and adjusting these to the increasing data availability and supplementing them with the use of HDM4 every few years to verify priorities and review different scenarios and their impact on future road conditions. Recent changes in the country have put the sustainability of the RAMS at risk, however, and it is not clear at this stage if MPW will continue supporting the RAMS and its development.

Performance-based road maintenance contracts. Afghanistan has extensive experience with piloting PBCs, although its application in recent years has been limited. Further development of the PBC approach in the future is recommended, determining the suitable contract size and duration for domestic contractors and developing a suitable set of performance standards and inspection methods that are easily applied in the context of Afghanistan and its complex security situation. The use of performance standards should also be extended to maintenance works carried out by the recently established *Qway-e-kar*. The use of locally based microenterprises for unpaved roads and off-carriageway maintenance has been very positive in other countries, and the experiences in Afghanistan should be reviewed and replicated where appropriate. This may include their involvement as nominated subcontractors under larger PBCs. Here too, recent developments in Afghanistan require future support to PBCs to be reviewed.

Azerbaijan

Road network. As of 2019, Azerbaijan has a total of 13,671 km of public roads,[7] including 1,911 km of magistral roads, 1,809 km of republican roads, and 9,950 km of collector and local roads. The condition of the road network has improved significantly over the past decades, with 94% of the magistral, 84% of the republican, and 47% of the local roads currently considered to be in good or fair condition. In terms of road activities, distinction is made between winter and summer maintenance (excluding pavement or structure repairs), current repairs (routine pavement and structure repairs), midterm repairs (periodic maintenance), capital repairs (rehabilitation), reconstruction (change to geometry), and new construction.

Institutional framework. From 1993 to 2003, responsibility for the public roads (magistral, republican, collector, and local roads) lay with the state joint stock company Azeravtoyol State Corporation that reported directly to the Cabinet of Ministers. In 2003, responsibility for the public road network was transferred to the newly established Ministry of Transport (MOT) and Azeravtoyol State Corporation was converted to the Road Transport Service Department (RTSD) under MOT. In 2007, RTSD was converted to an open joint stock company, Azeryolservis, under MOT and in 2015, Azeryolservis became independent from MOT as a state-owned company with a board of directors reporting directly to the Cabinet of Ministers. In 2016, its name was changed to Azeravtoyol, an open joint stock company, and subsequently in 2017, it was restructured to become a public legal entity Azeravtoyol State Agency.[8] Where road maintenance and repair used to be carried out by 63 district-based road maintenance units under Azeryolservis, this is now carried out by a total of 103 state-owned limited liability companies under Azeravtoyol, including 7 corridor-based motorway maintenance companies, 8 companies for maintenance of highways around Baku, 18 specialized road maintenance companies, 54 district-based road maintenance companies, 12 landscaping companies, and 4 other companies.

Financing. A Road Fund existed from 1994 until 2000, with revenues reaching $24 million in 2000. After the abolishment of the Road Fund in 2001, all revenues went to the central budget with annual budget allocations to the road sector. Road sector allocations initially dropped to $8 million in 2001 (including $6 million for road maintenance), but by 2006 the road sector allocation increased to $107 million (including $56 million for road maintenance). In 2007, the Road Fund was reinstated, with revenues for the road sector reaching $144 million in 2007 and $181 million in 2009 (including $80 million for road maintenance). In 2019, the revenue of the Road Fund reached $207 million, with spending on road maintenance reaching $177 million. The use of the Road Fund is limited to the operation, maintenance, and repair (current, midterm, and capital) of magistral, republican, local, and urban roads. Revenues from the Road Fund is increasingly spent on road maintenance and current and midterm repair, while rehabilitation tends to be financed separately from annual allocations from the central budget. Unused funding of the Road Fund is rolled over to the subsequent year.

An HDM4 analysis carried out in 2010 found that an optimal budget of $288 million was required for the first 5 years to address the maintenance backlog, followed by $136 million per year for the following 15 years, allowing 88% of the network to be brought to good or fair condition by 2030. The existing budget expenditure on maintenance and rehabilitation of $243 million in 2009 ($181 million from the Road Fund and $62 million

---

[7] This excludes the roads in the Nagorno–Karabach region and the Nakhchivan Automomous Republic.
[8] Azərbaycan Avtomobil Yolları Dövlət Agentliyi—State Agency for Azerbaijan Automobile Roads.

from the state budget) was largely in line with the identified needs. The study concluded that the amount of funding for maintenance and repair should be considered appropriate, but that allocation needed to be improved, giving higher priority to periodic maintenance and to roads in good to fair condition. With Road Fund revenue having since increased and a significant portion of the maintenance backlog in international and republican roads addressed in the past decade, this conclusion is expected to still be valid.

Road user charges. Until 2000, the Road Fund received earmarked funding from fuel tax (52% on petrol and 12% on diesel), corporate turnover tax, annual road license fee, new vehicle sales tax, and international transit fees. In 2001, the corporate turnover tax and the vehicle sales tax were abolished together with the Road Fund, with all remaining funding sources transferred to the central budget. In 2007, the Road Fund was reinstated, with the main revenue consisting of a "road tax" that is defined in the Tax Code. This road tax used to involve an annual vehicle tax based on engine size, but this was abolished for domestic vehicles in 2015 and replaced by a fuel tax on the wholesale price of locally produced fuel and the customs value of imported fuel ($0.011/liter).[9] The road tax also continues to include a vehicle tax for foreign vehicles[10] and a tax for foreign heavy trucks. Other revenues of the Road Fund include a simplified tax for transporters of passengers and goods (based on the number of seats or load capacity), an excise tax on imported vehicles, customs duties on imported vehicles, fees for international transport permits, fees for annual technical inspections of vehicles, and fines for oversized and overweight vehicles. With ADB support, the government is also introducing tolls on motorways, with tolling currently being piloted on a stretch of the M1 motorway.

Road Asset Management System

Development. RAMS development in Azerbaijan has been supported by several development partners over the past 2 decades under a range of different projects and programs. The World Bank has provided significant support under the different highway projects, with ADB providing complementary support under its road network development programs. The European Bank for Reconstruction and Development (EBRD) has also provided technical assistance in support of RAMS development. The support from development partners is still ongoing and is now focused on the institutionalization of the RAMS, including its integration into planning and budgeting procedures.

Data collection equipment. A set of survey equipment together with related vehicles were procured under the Second Highway Project. This includes a Greenwood laser profiler (to measure international roughness index and rutting), Road Doctor Camlink video-logging equipment, GPS receivers, a trailer-mounted Dynatest falling weight deflectometer (FWD), a Geophysical Survey Systems Incorporated ground penetrating radar (GPR), and 25 Metrocount mobile tube traffic counters. An accelerometer-based device for measuring road roughness was introduced at a later stage for use in roads in poor condition or unpaved surface.

Data collection. Data collection was carried out under the First Highway Project in 2002, but was not repeated in subsequent years, complicating the use of the database and the pavement management system (PMS) developed under the same project. Between 2010 and 2014, an extensive data collection exercise was undertaken as part of the Second Highway Project in 2006. Data collection was undertaken by a consultancy firm for 13,448 km of roads, involving surveying of a total 27,250 lane-kilometers of road using the equipment procured under the same project. Inventory data were collected for all public roads through a combination of

---

[9] This fuel tax is in addition to existing excise taxes on fuel that go to the state budget.

[10] The vehicle tax used to be applied to domestic vehicles as well but was limited to only foreign vehicles in 2015 when the fuel tax was introduced. The rate depends on the length of stay as well as the engine capacity (passenger cars), the seating capacity (buses), or the number of axles (trucks). Trucks pay an additional rate per kilometer traveled.

field surveys and post-processing of collected video and GPS data. The inventory data included GPS centerline, road length, road type, carriageway type, number of lanes, pavement type, shoulder type, pavement width, shoulder width, median width, and kerb stones. Detailed condition data were collected for all paved magistral and republican roads, and included roughness, rutting, cracking, transverse cracking, raveling, potholes, edge break, joint spalling, faulting, and drainage condition class. For the paved local roads, only a general condition classification was carried out, while for unpaved roads condition data were not collected. The roughness and rutting data were collected using the laser profiler, while surface distress data were obtained through post-processing of the collected video data (visual assessments). The condition survey was repeated for a total of 3 consecutive years between 2010 and 2013.

FWD data were collected for 4,952 km of paved roads in good to fair condition (6,706 km of data collection, some roads in two directions), and GPR data were collected for 7,981 km of roads (9,916 km of data collection, some roads in two directions). Over 1,100 bridges were also surveyed, including GPS location and bridge element condition classifications. Traffic data were collected for all magistral and republican roads through 1-week traffic counts in 249 locations using mobile traffic counters, while for the local roads, moving traffic counts were carried out based on the collected video data. The 1-week traffic counts were found to be less suitable in roads with high traffic volumes which would benefit from permanent traffic counters as the mobile traffic counts were found to be very inaccurate.

The data collection was extensive, and as a result expensive, costing $9.7 million in data collection alone, and an additional $0.8 million in survey equipment. A survey plan was prepared for execution by Azeryolservis in 2013 and 2014. This proposed GPS centerline, FWD, and GPR data collection were only carried out for new road sections. Annual surveys of 50% of magistral and republican roads and 35% of local roads would be carried out to collect video data (inventory and surface distress) and profilometer data (rutting and roughness). Bridge surveys would be carried out for 25% of bridges every year to update condition data. New traffic counts would be carried out in 50% of the counting stations each year. The survey plan also required that Azeryolservis road maintenance units report annually on the works performed, in order that this information may be updated in the RAMS database. This ambitious survey plan was not followed in practice, however, and subsequent projects have supported Azeryolservis and subsequently Azeravtoyol in the updating of some of the data.

Database. The first Highway Project (2001–2009) supported the development of a basic Microsoft (MS) Access database to store data collected under the project. The database included a function for exporting data for use in the HDM4 software. The ADB East–West Highway Improvement Project (2006–2010) carried out further improvements to the database and developed plans for upgrading and utilizing the RAMS. Although the MS Access database was supposed to be used for storing the data to be collected under the Second Highway Project (2006–2015), it was later decided to develop a new database for this purpose. The new database was completed in 2012 based on MS SQL Server. It includes inventory and condition data for roads and bridges together with photographs and videos. The database allows all collected data to be shown in tables, graphs, maps, and video or images, allowing easy review of the network and comparison of different features of the roads. The database allows different standard reports to be prepared and data to be exported to MS Excel, HDM4, or other software for further use, either through standard procedures or based on customized requirements. Data can also be processed and transformed before being imported into the database. The database introduced a location-referencing system based on nodes and road segments, but also allows all data to be presented in terms of road chainage in line with the previously existing system. The development of the database was followed by a 2-year support period, during which issues were corrected and the functionality was further improved. Although the database continues to be operational, it is only accessible in the central office of Azeravtoyol and is not accessible from the different road maintenance companies.

Data analysis. The World Bank first Highway Project supported the development of a PMS using the HDM4 software and data exported from the MS Access database developed under the same project. A 5-year highway management program was also prepared by the consultants. Although the project was successful in the installation of the database and the PMS as well as in training RTSD staff in their use, after completion of the project the PMS was no longer utilized by RTSD. Under the ADB East–West Highway Improvement Project, the calibration of the HDM4 software was updated to allow it to be used for prioritizing and planning pavement works.

Under the World Bank Second Highway Project, the developed database included comprehensive functions for data analysis using tables, graphs, and maps, as well as automated reporting functions and the ability to export data for further analysis. HDM4 was used by the consultants for preparing annual and multi-annual plans. The consultants also prepared a comprehensive Road Investment and Maintenance Master Plan using the data collected and the database developed under the same project.

Although the RAMS continues to be operational, the data analysis tools have not yet been integrated into the annual planning and budgeting procedures. Costing and budget allocation continue to be on the basis of a decree from 2006 setting the average costs per kilometer for roads depending on their technical class (ranging from $10,000/km/year for gravel roads to $55,000/km/year for the highest-class asphalt concrete roads). Planning and budgeting do not take proper account of the road conditions and related treatment needs, nor do they take account of traffic volumes and other factors determining the importance of different roads.

Performance-Based Road Maintenance Contracts

In the year 2000, all roadworks were still carried out through force account by the Azeravtoyol State Corporation. In the early 2000s, state-owned construction companies were privatized, and construction and rehabilitation works started to be contracted out on a competitive basis. Periodic maintenance is currently also mostly contracted out. Routine and winter maintenance and current repairs continued to be carried out by the road maintenance units under RTSD and Azeryolservis, and are currently carried out by 103 corporatized road maintenance companies under Azeravtoyol. Private sector participation in road maintenance and current repairs is extremely limited. The EBRD prepared a PBC model contract document for competitive tendering in 2018, but tendering has not yet taken place due to the limited interest of Azeravtoyol in private sector involvement. With World Bank support, performance-based, service-level agreements (SLAs) are being piloted with some of the motorway maintenance companies under Azeravtoyol.

Pilots. The World Bank Third Highway Project supported the establishment of seven motorway maintenance companies responsible for the routine, winter and emergency maintenance, and current repairs of the country's motorway network. The project has also been supporting the introduction of performance-based SLAs with three of these motorway maintenance companies, covering 774 km of the M2 and M4 motorways. These are direct contracts with subsidiaries of Azeravtoyol against negotiated prices and do not involve competitive bidding. The cost for one of the contracts covering 264 km of the M2 amounts to $4 million per year for the performance-based routine summer and winter maintenance, equivalent to $15,000/km/year. Activities include monitoring, cleaning and maintenance of pavement, drainage, signalization, signage, and winter maintenance, which are paid through monthly lump-sum payments against compliance with the performance standards. Occasional emergency repair works are also included, but are paid through additional payments based on completed volumes using unit rates. Periodic maintenance is not included in the scope of the contracts with the motorway maintenance companies and is contracted out separately on a competitive basis. Under the SLAs, payments are to be made on a performance basis by Azeravtoyol using financing from the Road Fund, with deductions in case of poor performance. After the start of the contracts in January 2019, the calculation of the payments and penalties (deductions) was initially not in accordance with the contract agreements, but was resolved toward the end of 2020 and the payments and deductions now follow the compliance with the service levels.

Performance standards. The SLAs include a set of 61 different performance standards related to (i) roadway and other paved areas; (ii) vegetation control; (iii) road surface cleanliness and safety; (iv) signage and safety devices; (v) electrical, mechanical, and systems maintenance; (vi) winter maintenance; (vii) drainage; (viii) earthworks and roadside; (ix) structures; (x) traffic incidence response; and (xi) buildings and miscellaneous facilities maintenance. Most performance standards are clearly defined and are easily verified (e.g., no potholes allowed that are greater than 20 centimeters [cm] in any dimension, less than 10% of cross section of culvert is blocked), although some standards remain subjective (e.g., no safety hazards). All performance standards are linked to response times, within which an identified defect has to be corrected. The response times start counting from the moment the defect is identified by either the motorway maintenance company or Azeravtoyol. Most response times are for 7, 14, or 28 days, providing sufficient time to correct the defect. A large number of performance standards, however, involve short response times of only 1–48 hours. These short response times are related to sudden events (e.g., immobilized vehicles, snowfall), but also include several predictable events (e.g., potholes). Winter performance standards especially have very strict response times, requiring any snow over 5 cm in depth to be removed within 2 hours with subsequent cleaning every 2 hours, and all snow on the road surface, shoulders, and drainage to be removed within 48 hours after snow stops. The short response times are also applied to the buildings of the motorway maintenance companies, with any heating, electrical, and plumbing problems to be repaired and overflowing garbage cans to be cleaned within 24 hours.

Deductions are applied to the monthly lump-sum payments only if the identified defects have not been corrected within the defined response time. The deduction is calculated based on the time taken to repair the identified defect beyond the response time, with most performance standards involving one penalty point for each extra day until the defect has been repaired. This means that overflowing garbage cans at the motorway maintenance company buildings result in the same penalty as large potholes or blocked culverts on the motorway. Only in the case of obstacles, safety hazards, or snow and/or ice on the road is the deduction higher with one penalty point applied for every hour beyond the response time. Each penalty point is equivalent to 0.1% of the monthly payment and appears to be applied to each 1 km section, although this is not clear from the SLA documents. Even though the penalty points accumulate over time, it would require 1,000 penalty points to lose the entire monthly payment. Most performance standards accumulate at most 31 penalty points in a month, equivalent to only 3% of the monthly payment. It would therefore require noncompliance with over half the performance standards for the entire month before the deduction equals the full monthly payment.

The use of response times means that the motorway maintenance companies are inclined to await the inspections before carrying out any repairs, undermining the proactive nature of PBCs. But they also mean that inspections need to be carried out by Azeravtoyol on a more or less continuous basis to identify any defects and to check that previously identified defects have been corrected, increasing the management burden. The SLA documents are not clear on the frequency and procedures for inspections, except that these occur at least once a month and that subsequent payments are made at the end of the following month (in light of the applied response times to correct any identified defects).

Way Forward

Road asset management system. Azerbaijan has a very comprehensive database, and extensive data collection has been carried out in the past. The regular updating of the data is proving to be a challenge for Azeravtoyol, and continued support will be required to institutionalize the collection of data for a portion of the road network each year, including a dedicated budget to cover the costs of doing so. Adjustments to the data collection requirements may be required to reduce the time and costs involved in the collection and subsequent processing of the data. Although the database has an extensive functionality, including the option to export data to other software such as HDM4 for further analysis in support of planning and budgeting, this is not yet taking place on a regular basis. Here too, there is a lack of institutionalization of the data analysis, and a lack of integration into existing planning and budgeting procedures. The focus should now be on creating ownership of the RAMS and supporting amendments to the existing institutional and normative framework as well as to the RAMS itself to allow for its improved integration. This will require the amendment of certain norms and procedures to be more in line with the RAMS and the algorithms used for costing and prioritization. This may also require the introduction of more basic and transparent algorithms for costing and prioritization (e.g., decision matrix) that can be better aligned with existing practices.

Performance-based road maintenance contracts. Azerbaijan has already progressed considerably in the introduction of a contractual relationship with its subsidiary road maintenance companies. Performance-based SLAs are being piloted with a number of these companies and will likely be replicated with other companies in the future. Further improvement of the performance standards and deduction mechanisms is required, ensuring that deduction levels reflect the seriousness of the defect concerned and reducing the dependency on response times and related inspection visits. The SLAs still depend on directly awarded and negotiated contracts with subsidiary companies, and as such will lack some of the benefits of tendered contracts. A next step will be to pilot PBCs with the involvement of private sector companies through competitive tendering. The state-owned road maintenance companies should be able to participate in these tenders, as long as they do so under the same conditions as the private sector companies. The introduction of competitive bidding for PBCs may also be combined with a gradual privatization of the state-owned road maintenance companies, as has happened in various other countries. This will need to take into account the risks of staff redundancies in the state-owned companies, and will need to introduce acceptable options for these staff members, building on similar experiences from other countries.

People's Republic of China

Road network. According to the *China Highway and Waterway Transport Statistics Yearbook*, the People's Republic of China (PRC) had a road network encompassing nearly 5.20 million km in 2020. This includes 160,980 km of expressways, 123,101 km of class I highways (four lanes or more), 418,300 km of class II highways (two lanes), 4.24 million km of class III and class IV roads (one or two lanes), and 253,632 km of underclass roads.[11] According to the Technical Specifications for Highway Maintenance (JTG H10-2009), distinction is made between minor maintenance (frequent preventive maintenance and repair of the road and its facilities), medium repair (regular repair and strengthening of general damaged road sections and facilities), and major repair (comprehensive repairs for major damages to the road and facilities). These maintenance and repair activities aim to bring the road back to the original technical standard and are distinguished from (re)construction, where the technical standard is changed to meet growing traffic volumes and load requirements.

Institutional framework. The Ministry of Transport (MOT) has delegated responsibility for the management of the trunk road network of national and provincial roads, as well as some major county roads to the departments of transport and their highway administration bureaus in the 27 provinces and four major municipalities. In some provinces, responsibility for these trunk roads has been delegated to the prefecture (municipality) level. Responsibility for the lower-level county, township, and village rural roads and urban streets lies with the prefectures and the underlying county transport bureaus. In most provinces, minor maintenance in the national and provincial highways is carried out by county-level maintenance stations under the highway administration bureaus. Medium and major repairs tend to be tendered competitively, but are often awarded to the business units of the prefecture-level maintenance divisions under the highway administration bureaus. In the case of expressways, these tend to fall under the provincial departments of transport and are generally managed by public companies set up specifically for this purpose.

Financing. Financing for the trunk road network comes primarily from government fiscal allocations. Especially in the poorer western provinces, financing is highly dependent on road user charge revenues collected at central or provincial levels, complemented by annual allocations from national and provincial budgets that vary from 1 year to the next.[12] The most important source of funding is the vehicle purchase tax, which is dedicated to road construction and rehabilitation. The ADB study on *Reforming the Financing System for the Road Sector in the People's Republic of China* (2015) estimated that between 2011 and 2020, the vehicle purchase tax would provide approximately $600 billion in revenue. Fuel tax allocations make up another important source of revenue collected at the central level, with allocations to the provinces used mainly for road maintenance and daily operations. Revenue from the fuel tax allocated to the road sector was estimated to amount to approximately $300 billion for the period 2011–2020. Remaining revenues for the trunk road network come from central and provincial budget allocations, as well as other locally collected road user charge revenues such as traffic fines. The $900 billion in revenue estimated to be collected from the vehicle purchase tax and the fuel tax in the 10-year period from 2011 to 2020, should be compared to the estimated investment of $1,000 billion for planned road development and $600 billion for road maintenance for the same period. Where there is a parallel demand for road development, road maintenance tends to get underfunded.

---

[11] These roads do not comply with the minimum standards of class IV roads.
[12] This mainly involves financing for road improvements.

Road user charges. The PRC collects several road user charges, some at central level and others at provincial or local level. One of the most important is the 10% vehicle purchase tax that is allocated to the different provinces and is dedicated to road development (construction and rehabilitation). In 2019, the revenue collected from the vehicle purchase tax amounted to $57.4 billion. Up to 2009, provinces used to collect tolls and a vehicle maintenance fee that was used to finance road maintenance activities. In 2009, a fuel tax reform was implemented, increasing the fuel tax from $0.04 to $0.21 per liter for unleaded gasoline and from $0.02 to $0.12 per liter for diesel.[13] Over the years, the rates have further increased to $0.23 per liter for gasoline and $0.18 per liter for diesel. Locally collected vehicle maintenance fees and all tolls on class II highways were abolished in 2009 and replaced by fuel tax revenue allocations from the central government. In 2019, the revenue from the fuel tax amounted to approximately $80 billion, although not all fuel tax revenues were allocated to the road sector. The fuel tax currently forms an important source of funding for highway maintenance, with approximately 20% of fuel tax revenues allocated to the provincial highway administration bureaus for the maintenance of national and provincial highways. Additional allocations are made to the prefecture level to subsidize the maintenance of rural roads, based on 2007 revenue levels from vehicle maintenance fees and class II highway tolls, adjusted for inflation. Tolls continue to be collected on class I highways and expressways, but mainly serve to finance the maintenance of these highways and expressways and the repayment of the investments in their construction.

Road Asset Management System

Development. In 2007, Highway Performance Assessment Standards (JTG H20-2007) were issued by the MOT, introducing a set of indicators for defining road conditions. This is centered around the maintenance quality index (MQI) which is based on four indicators: the pavement quality index (PQI); the subgrade condition index (SCI); the bridge, tunnel and culvert condition index (BCI); and the traffic facility condition index (TCI).[14] The most important of these indicators is the PQI. According to the latest update of the standards in 2018 (JTG 5210-2018), the PQI is now based on seven sub-indicators: pavement condition index (PCI), riding quality index (RQI), rutting depth index (RDI), pavement bump index (PBI), pavement surface wearing index (PWI), skid resistance index (SRI), and pavement structure strength index (PSSI).[15] For class I highways and expressways, all indicators are used, while for paved roads of class II and lower, only the PCI and RQI are used, and for unpaved roads only the PCI is used. Detailed calculation formulas exist for each of the indicators, which are set up in such a way that the scoring ranges from 0 to 100, both for the individual and compound indicators. The score is translated into condition classes, whereby a score of 90 or higher is rated excellent, 80–90 is good, 70–80 is fair, 60–70 is poor, and below 50 is very poor (these condition classes are applied to all indicators).

In the late 1980s, the PRC developed the China Pavement Management System (CPMS). Although the government intended to roll out the CPMS to all provinces by the year 2000, this faced problems in software development and alignment with existing standards. Together with the introduction of the Highway Performance Assessment Standards, the CPMS was also updated and formally renamed the CPMS National and Provincial Trunk Highway and Expressway Assets Management System. The enhanced CPMS is fully in line with the highway performance assessment standards and provides a comprehensive system for monitoring and predicting network performance. The fact that the CPMS is in line with existing assessment standards has resulted in the CPMS software becoming the standard for road asset management in the PRC.

---

[13] According to the ADB report on *Financing Road Construction and Maintenance after Fuel Tax Reform* (TA 7456-PRC), the tax for gasoline increased from CNY0.2/liter to CNY1.4/liter, and for diesel from CNY0.1/liter to CNY0.8/liter.

[14] MQI = 0.70*PQI + 0.08*SCI + 0.12*BCI + 0.10*TCI.

[15] For roads of class II and lower, PQI = 0.6*PCI + 0.4*RQI. For asphalt class I highways and expressways, PQI = 0.35*PCI + 0.4*RQI + 0.15*RDI + 0.10*SRI. For concrete class I highways and expressways, PQI = 0.50*PCI + 0.4*RQI + 0.10*SRI.

Data collection equipment. In line with the introduction of the Highway Performance Assessment Standards and upgrading the CPMS, the PRC also developed road network survey equipment that allows data regarding potholes, roughness, rutting, cracking, geometry, and video to be collected at normal driving speed. Data are used to calculate the PCI, RQI, and RDI, forming the main inputs for the calculation of the PQI and the MQI. Although the central government recommends the use of road network survey vehicles for collecting data, the approach also allows data to be collected manually. These road network survey vehicles have increasingly become available in all provinces to collect the necessary data, at least with the provincial highway administration bureaus (for rural roads, data are still often collected manually).

In Yunnan province, for example, the Yunnan Highway Administration Bureau (YHAB) purchased a road network survey vehicle in 2009, and provided this to the Yunnan Province Highway Research Institute. A second similar vehicle was rented and later purchased by the YHAB to respond to the objective of surveying 15,000 km of roads per year, covering multiple lanes (in 2009, a total of 46,500 lane–kilometers were surveyed). The road network survey vehicles can collect data on pavement damages, roughness, rutting, road structure, road furniture, and road signs. With support from the World Bank, the two road network survey vehicles were upgraded in 2016 and a third vehicle was procured considering the expected increase in annual surveys to cover 39,000 km of roads. An automated deflectometer for measuring pavement strength was also provided, as were nine bridge and eight tunnel inspection vehicles to inspect the 13,205 bridges and 783 tunnels in the province. To complement the existing 34 traffic counters and cover the 1,987 traffic counting locations in the province, 63 permanent traffic counters and 80 portable traffic counters were provided.

Data collection. The indicators introduced by the Highway Performance Assessment Standards form the basis for road data collection. The most important of these indicators are the PCI and RQI. For roads of class II and lower, only the PCI and RQI are considered in calculating the PQI (for unpaved roads, only the PCI is used), and with 70% weight, the PQI is by far the most important indicator in calculating the overall MQI. The PCI looks at various types of surface distress: crocodile, block, longitudinal and transverse cracking, potholes, raveling, depressions, rutting, shoving, bleeding, and patching. The PCI is calculated by adding up the areas affected by each type of distress multiplied by a weighting factor, and dividing this by the total road surface area. The weighting factor ranges from 0.6 to 1.0 and depends on the type and severity of the surface distress. The area affected by surface distress can be determined manually, but is increasingly measured automatically using road network survey vehicles. The RQI depends on the International Roughness Index (IRI), but applies a specific formula to achieve a range of 0–100 in the same way as for the other indicators.[16] Roughness can be measured using a 3-meter straight edge, but is generally measured using laser profilometers.

All road management entities at the different levels of government are required to report on the condition of their national, provincial, county, and lower-level roads on an annual basis, and data collection is therefore carried out each year (all roads are surveyed at least once every 2 years). Since 2008, the collected data are reported in annual highway maintenance statistics reports together with other statistical data on the road network and its management. Although data are aggregated and stored for 1 km road sections, it is reported either by road (average for the full road) or by maintenance division (average for all the roads managed by the maintenance division). The use of averages does not allow for proper analysis of road network conditions, as it is not clear how the average condition is distributed over the network.

Most provinces have specialized units responsible for data collection, which also have the necessary road network survey vehicles and other survey equipment (these are often located within the research institutes under the provincial departments of transport). For the rural roads managed by the county transport bureaus,

---

[16]   RQI = $100 / (1 + 0.0185 * e^{0.58*IRI})$ for roads class II or below (for expressways and class I roads the factors are slightly different). An IRI of 3 is equivalent to an RQI of 90 (excellent), while an IRI of 7 is equivalent to an RQI of 50 (very poor).

data collection is generally done manually at fixed times each year by in-house staff. To increase objectivity, assessments are carried out by staff from other counties that are not directly involved in the maintenance of the roads.

Database. The original CPMS consisted of three modules: road database, network-level maintenance management system, and project-level maintenance management system. It had been adopted by several provinces across the country, often with support from development partner projects. However, it had high data requirements, poor user interface, limited network analytical functionality, and could only be applied to bituminous pavements (a large portion of the paved network has a cement concrete surface). The enhanced CPMS National and Provincial Trunk Highway and Expressway Assets Management System developed in the 2000s fully incorporates the different indicators of the highway performance assessment standards. It includes 11 modules classified into four categories: asset database, demand analysis and planning tools, maintenance implementation management tools, and monitoring and reporting tools. The development of road network survey vehicles has facilitated the collection of the necessary data, and the enhanced CPMS is now used across all provinces in the country. In some provinces, the CPMS is only used for road network monitoring and preparing annual reports, and the analytical modules have not been procured or are not used. However, development partners have been supporting the introduction of additional CPMS modules, specifically through the procurement of the system and survey equipment as well as through the training of staff.

The asset database module is used to store the data collected during the yearly highway performance assessments and to prepare the annual highway maintenance statistics reports. The enhanced CPMS database allows inventory and condition data to be stored regarding (i) road geometry; (ii) subgrade, shoulder slope, and ditch; (iii) pavement; (iv) bridges and tunnels; (v) road furniture and signage; (vi) vegetation and greening; and (vii) the right-of-way, as well as data on traffic volumes and composition. Where automated survey vehicles are used, the data can be directly uploaded into the database. The CPMS does not include a comprehensive geographic information system (GIS) module, however, and this is generally added on to the CPMS using a separate software.

Data analysis. In the original CPMS. the network-level maintenance management module was quite basic. This looked at the score for PCI, RQI, and PSSI to determine whether a road required minor maintenance, medium repair, or major repair. For instance, a road with a PCI score of 75 or higher and an RQI score of 90 or higher would require only routine maintenance. If the RQI score was lower, medium repair would be planned. If the PCI score was lower, the PSSI score would determine whether medium repair or major repair was required. This was found by the provinces to not properly reflect the actual needs and priorities and the planning module was not widely used.

In the enhanced CPMS, the analytical module for pavements has been significantly improved and is now able to optimize budget allocations under funding constraints and predict future pavement conditions, allowing optimized and prioritized annual and multi-annual work programs to be prepared. The modules for bridges and tunnels remain simple, and mainly serve to manage inspection data and provide overall condition ratings in compliance with national standards.

Performance-Based Road Maintenance Contracts

Pilots. Several PBC pilots have been carried out in different provinces with development partner support. This report describes the pilots in Yunnan and Anhui provinces as examples, although other experiences exist in some other provinces. In Yunnan province, the ADB-funded Yunnan Integrated Road Network Development Project included several pilots with performance-based rural road maintenance carried out by road maintenance groups, which comprised of women living near the roads to be maintained. They were paid a fixed amount per month against compliance with a set of basic performance standards. Based on the initial pilot in 2012, the approach was replicated with government funding in 650 km of township roads for the 3-year period between 2013 and 2016.

The ADB-funded Yunnan Sustainable Road Maintenance Project included two PBC pilots. The first involved a 5-year output- and performance-based road contract (OPRC) for 57 km of national highway that was tendered competitively and awarded in 2015. This included initial medium repair (17 km) and major repair (40 km) paid on an output basis, followed by 5 years of performance-based minor maintenance. The contract amount was $9.7 million, equivalent to an average of $33,250/km/year inclusive of initial medium and major repair, and was awarded to a Chinese company located in Yunnan province. Staff from the maintenance stations traditionally responsible for the minor maintenance of the highway concerned was transferred to other areas or other positions within the YHAB.

The second pilot involved a county-level road maintenance unit under the YHAB, introducing a service-level agreement (SLA) for performance-based routine maintenance of 107 km of national and provincial highways. The SLA was signed between YHAB's prefecture-level maintenance division and YHAB's county-level maintenance unit in 2014 with a duration of 3 years. Although the project was initially designed to cost $5 million, initial repair costs turned out to be more expensive and the total cost increased to $8 million, equivalent to an average cost of $25,000/km/year inclusive of initial repair works (this did not include staff and equipment replacement costs of the maintenance unit).

The World Bank-funded Anhui Highway Rehabilitation and Improvement Project (2008–2012) included the piloting of two PBCs in provincial highway sections of 60 km each over a period of 2 years from 2009–2011. The pilot was budgeted at $1.6 million, but ended up costing $4.8 million (equivalent to $20,000/km/year) due to extensive medium repairs (34.5 km) and major repairs (21.4 km). The pavement condition was significantly improved over the contract period, with the PCI improving from an average of 67 (poor) to 89 (good). The overall condition similarly improved from an MQI of 72 (fair) to 91 (excellent). The pilot was found to be very successful by the Anhui Highway Administration Bureau, although they considered the contracted road length to be too small and the duration too short, limiting the potential for assessing the performance of the contractors and their ability to reduce costs through improved maintenance management.

In the subsequent Anhui Road Maintenance Innovation and Demonstration Project (2017–2023), a total 776 km of roads is to be put under eight OPRCs of 5 years duration. Contract lengths vary from 80–160 km, with one smaller contract of 50 km (this includes a greater proportion of major repairs). Six of the eight contracts include performance-based minor maintenance combined with medium and major repairs paid on an output basis, with the contract defining the maximum annual repair lengths (the contractor is free to decide the location of these repairs). The other two contracts included only performance-based routine maintenance. So far, only one contract has been awarded and three more are under evaluation. The awarded contract is for 88.5 km of national highway at a cost of $12.8 million, including 5 years of minor maintenance as well as necessary medium and major repairs, equivalent to $29,000/km/year on average. All contracts are tendered using national competitive bidding. Staff from the maintenance stations traditionally responsible for the minor maintenance of the different highways concerned was transferred to other areas or other positions within the Anhui Highway Administration Bureau.

Performance standards. Under the Yunnan Sustainable Road Maintenance Project, the tendered OPRC made use of the government's Model Bidding Documents for Procurement of Civil Works under National Competitive Bidding. In the case of the PBC pilot with YHAB's county-level maintenance unit, use was made of a basic two-page SLA based on existing agreements used in the YHAB. The contract documents included a Bill of Quantities for the initial repair works, a fixed monthly lump sum for the maintenance works, and a provisional sum for emergency works. The specifications of the performance-based maintenance were defined in the Employer's Requirements that formed part of the contract documents. This included a set of specific performance standards, consisting of five standards for the right-of-way and drains; 11 standards for the carriageway; three standards for tunnels, culverts, and bridges; and six standards for road furniture and signage. The performance standards use clear thresholds that facilitate objective measurements (e.g., maximum diameter and depth of potholes, maximum number of smaller potholes per 1 km section). Monthly inspections assess the compliance with the performance standards. Response times are not included, and noncompliance with the performance standards at the time of inspection leads to immediate deductions to the monthly payment. Deduction percentages are defined for each performance standard, ranging from 10% for kilometer posts to 50% for potholes and cracking. These are applied for each 1 km section that is found to be noncompliant, and multiplied by the monthly payment per kilometer to calculate the total payment deduction. If defects are not corrected before the next inspection, the deductions are doubled. A separate road usability standard requires that the road is open to traffic at all times, with closures limited to a maximum of 6 hours, and penalties equivalent to 20% of the full monthly lump-sum payment for every 24 hours (or portion thereof) that the closure is not resolved. Timely submission of reports and compliance with plans is also subject to payment deductions of 5% of the monthly lump sum. Noncompliance with the different performance standards is recorded in a monthly inspection report that also serves to calculate any possible deduction to the monthly payment. The same performance standards and deduction calculations were applied in both the OPRC and the SLA pilot, facilitating comparison of performance.

In the Anhui Road Maintenance Innovation and Demonstration Project, the performance standards are based on the highway performance assessment standards. The contractors are required to achieve a score of at least 85 for each of the main indicators (mainly PCI and RQI, and the resulting PQI and MQI) for each 1 km section in case of national roads, and a score of 80 in the case of provincial roads. The average PQI of the project roads was 60 at the start of the project, but significant improvements were achieved through the medium and major repairs, with the performance-based minor maintenance mainly aimed at preserving these improvements for the full 5-year contract period. Although the use of the highway performance assessment standards allows proper linkage with the annual road condition assessments, these indicators are not very suitable for use in monthly inspections as they require various measurements along the full contracted road length.

Way Forward

Road asset management system. With over 5 million km of roads to be managed, the PRC has developed a detailed system of highway performance assessment standards and complemented this with its own RAMS software in the form of the CPMS, data collection equipment, and annual reporting requirements. The annual reporting requirements introduced the practice of annual data collection, even when this was not yet used in planning. The availability of road network data together with the gradual rollout of the CPMS has enabled the provinces to prepare evidence-based annual plans. Further strengthening data collection and the use of the CPMS for planning will continue to be required, especially in the poorer western provinces and for the lower-level roads. The functionality of the CPMS may continue to be improved, for instance adding a comprehensive GIS mapping module and expanding the functionality of bridge and tunnel maintenance planning.

Performance-based road maintenance contracts. The PRC currently has experience with PBC pilots in various provinces, applying different modalities. Some involve tendered contracts with domestic companies, while others involve in-house maintenance units. Different performance standards and deduction calculations have been used. A detailed evaluation of the PBC pilots is required in order to collect the lessons learned and identify the best practices, so these may be used in the further replication of the approach in those circumstances where it is considered suitable. This will also need to include a cost comparison, especially between the traditional approach and the performance-based approach, and between in-house and outsourced modalities. Such a cost comparison should take account of all costs, including salary, pension, and equipment costs that may not be covered under the contract or agreement itself.

Georgia

Road network. As of 2019, Georgia has 7,010 km of main roads, including 1,564 km of international roads and 5,446 km of secondary roads. In terms of road activities, distinction is made between routine maintenance, periodic maintenance, rehabilitation, and (re)construction.

Institutional framework. The main roads are managed by the Roads Department under the Ministry of Regional Development and Infrastructure. RAMS operation and planning are carried out by the planning and operations section under the Roads Department. Former in-house implementation units under the Roads Department were privatized in 1999 and all construction, rehabilitation, and maintenance are contracted out to the private sector. Government-funded routine maintenance is contracted under 3-year contracts covering 24 zones, with an additional four contracts covering tunnels and bridges. The 28 contracts have been awarded to 24 different contractors. Periodic maintenance is contracted separately. Management of government-funded works and works funded by development partners is done by separate divisions under the Roads Department.

Financing. Road sector financing is through annual allocations from the state budget. Financing has increased significantly in the past 10 years, from $150 million in 2008 to over $340 million in 2018, most of which is allocated to road rehabilitation and reconstruction. Most of this increase is related to development partner funding that increased from $25 million in 2008 to $150 million in 2018, gradually shifting from basic rehabilitation to complex upgrading of international and secondary roads. Apart from some PBC pilots, all maintenance is funded by the government. Maintenance funding has increased from $15 million in 2008 to $35 million in 2018, in line with the general road sector budget increase. Budget allocations to road maintenance and rehabilitation are largely in line with requirements as identified by the Roads Department in their annual planning, as determined using the RAMS.

Road user charges. Georgia collects several road user charges. The fuel excise tax more than doubled in 2017, increasing to $150 per ton for petrol ($0.14 per liter) and $125 per ton for diesel ($0.11 per liter). This provided a revenue of $345 million in 2018, forming one-third of the total excise tax revenue in Georgia and equivalent to the total road sector budget in that same year. A vehicle excise tax of $0.45–$0.75 per cubic centimeter of engine capacity is collected, whereby older vehicles are required to pay a higher excise tax. With just under 200,000 vehicles newly registered in 2019 (mostly second-hand vehicles), the revenue from the vehicle excise tax is in the order of $150 million to $200 million. However, none of this road user charge revenue is earmarked for the road sector and it goes straight to the general budget.

Road Asset Management System

Development. The Roads Department has been developing its RAMS since 2007, with support provided by the World Bank and ADB under a series of projects. Pavement condition data are collected every year for the international and secondary roads, and the RAMS forms the basis for annual and 5-year planning. The RAMS is managed by the RAMS unit under the Roads Department, consisting of three staff positions. The RAMS unit used to be a separate section, but was merged with the planning and operation section in 2018, resulting in a reduction of staff numbers. The operation of the RAMS was initially supported by a local RAMS specialist contracted under various projects, but this is now fully carried out by the RAMS unit. The local consultant continues to support the further development of the RAMS. The Roads Department is currently aiming to expand the data collection to include detailed inventory data and improve the functionality of the RAMS database.

Data collection equipment. In 2007, the Roads Department purchased a Road Measurement Data Acquisition System (ROMDAS) road survey equipment under the Second Secondary and Local Road Project. This included a vehicle with two laser profilometers (roughness), video cameras, distance measuring instrument (chainage), and a GPS receiver. This was expanded in 2017 under the Secondary Road Asset Management Project to include a geometry module and a 360-degree camera as the basis for the International Road Assessment Program (iRAP) for carrying out road safety assessments of international and secondary roads. Traffic data are collected using a combination of fixed and mobile traffic counters (laser, radar, and tube systems). This equipment belongs to the routine maintenance contractors that are required to collect the traffic data as part of their contract. The Roads Department currently has two mobile traffic counters and is keen to purchase additional units. It is currently considering expanding the data collection equipment to allow a detailed road inventory to be carried out to support the update of the road passport data. This will likely include a system for automated surface distress measurements to reduce the time involved in visual assessments, as well as a system for automated inventory data collection (laser- or video-based).

Data collection. GPS and video data have been collected for 1,225 km of international roads and 4,612 km of secondary roads managed by the Roads Department, as well as 9,000 km of local roads under the responsibility of the municipalities. Roughness data (IRI) has been collected for all international and secondary roads and is available for all years since 2007. International roads are surveyed each year and secondary roads at least once every 2 years (approximately 70% each year). Traffic data are collected three times a year in 48-hour counts using mobile traffic counters in approximately 200 locations and is available for the period since 2006. The traffic counts used to be carried out by Roads Department staff, but are now delegated to the routine maintenance contractors who are required to purchase the mobile traffic counters under their 3-year maintenance contracts. Inventory data collection has been limited to pavement data (surface type, width, number of lanes, etc.). For large structures (bridges and tunnels), location data and photographs have been collected. Data collection is carried out each year by the RAMS unit using the ROMDAS survey vehicle. The Roads Department is currently aiming to collect more detailed inventory data in line with its standards on road passports. The detailed inventory data collection will be carried out by consultants with development partner financing.

Database. The Roads Department used to make use of a simple MS Excel-based system that was developed in 2004. This was upgraded in 2007 and now makes use of ESRI ArcGIS as the basis for its GIS mapping as well as for the database for all collected data. ArcGIS is used as a standard across government and staff are well versed in its use. As the Roads Department has the required skills and experience, the software is suitable as the basis for the RAMS. However, the current ArcGIS Desktop version does not allow for multiple users or remote access, and as a result the GIS database files are shared back and forth on USB drives and changes are updated in the main file as necessary. This is causing issues with data security and backups. The ArcGIS Desktop version is also limited in its functionality for combining GPS coordinates with a linear referencing system (chainage), preparing strip maps, and other functions related to road management. The Roads Department is therefore planning to upgrade to ESRI ArcGIS Enterprise with the roads and highways extension. This will include a web-based interface to allow remote access for public and internal use, and includes functions for the preparation of standard and customizable reports and the export of data to common spreadsheet formats and for use in HDM4.

Data analysis. Data analysis is carried out using HDM4 based on data exported from the RAMS database. The Roads Department has developed a software application to generate homogeneous road sections (road class, width, traffic volume, surface type, roughness class, rutting, deflection, friction, and climate zone) based on the RAMS data. The homogeneous sections are subsequently imported into HDM4 to carry out a program analysis. The results of the HDM4 analysis are imported back into the RAMS database and visualized using the GIS mapping. The HDM4 results are subsequently subject to a multi-criteria analysis to prepare the final annual and 5-year plans. This multi-criteria analysis includes aspects related to population density, connectivity, poverty, and economic activities in the area served by each road. The final plans tend to coincide with the HDM4 analysis for about 80%, with the other criteria causing some changes. The annual plans provide the basis for annual budget requests, with budget allocations largely following the requests made.

Performance-Based Road Maintenance Contracts

Pilots. Under the World Bank-financed Second Secondary and Local Roads Project, the Roads Department carried out an output- and performance-based road contract (OPRC) pilot in the Kakheti region covering 117 km of secondary roads and running from 2016 to 2021. The contract included 37.5 km of rehabilitation that was paid on a lump-sum basis according to the length of completed road, complying with the geometry, roughness, and pavement strength requirements defined in the contract. Periodic routine and winter maintenance are included, as necessary, for the full 117 km for a period of 5 years and are paid on a performance basis. A provisional sum is included to cover possible emergency maintenance needs and paid according to unit rates. The roads involved run through flat terrain with relatively light snowfall to facilitate the piloting. Design of the rehabilitation and periodic maintenance works was carried out by the contractor. Initial tendering of the contract was canceled due to very high bid prices received (over three times the cost estimate). To a large extent this was due to risk pricing related to the maintenance of unpaved roads and bridges. As a result, the contract scope was adjusted to exclude gravel roads and bridges, reducing the overall length of roads from 225 km to 117 km in order to fit the project budget. A second tendering was more successful and resulted in the contract award to a joint venture of Georgian contractors[17] for a contract amount of $16.7 million (GEL40.5 million). The output-based rehabilitation forms 80% of the contract price, with the remaining 20% covering the routine and winter maintenance costs, averaging approximately $5,200/km/year (GEL12,500/km/year). The contract is functioning well, but required several contract variations in the first years.

The subsequent World Bank-funded Secondary Road Asset Management Project was planning to pilot a second OPRC contract in Guria Region covering 240 km of secondary roads over the period 2020–2025. The contract included 68 km of rehabilitation (55% of the contract price) paid according to unit rates, 107 km of periodic maintenance paid on a lump-sum basis[18] (25% of the contract price), and routine and winter maintenance for the full road length paid on a performance basis (20% of the contract price). An additional provisional sum was included to cover emergency maintenance (7% on top of the contract price). Based on the experience from the first pilot, it was decided to only keep the performance-based payments for the routine and winter maintenance, but to apply a lump-sum payment for periodic maintenance. The rehabilitation design was carried out by a design consultant, but periodic maintenance design was to be done by the contractor. The terrain is steeper and has more snowfall, making the contract more challenging. The contract duration was for only 5 years, including time spent on the rehabilitation and periodic maintenance works under the OPRC contracts. Bidders included Georgian and Chinese contractors, but due to high bid prices, the contract was not awarded, and the tender was canceled. Due to the coronavirus disease (COVID-19), it was decided not to carry out further amendments to the bidding documents to improve understanding by bidders, and instead a 62 km section of the proposed road was contracted using traditional output-based contracts covering only rehabilitation works.

---

[17] Two other qualified bids were received from international contractors.
[18] The pavement needs to comply with maximum roughness and minimum strength thresholds specified in the contract.

A third PBC was planned under the ADB-funded Batumi Bypass Road Project. This would cover 140 km of roads in Mtskheta–Tianeti region that were recently rehabilitated, with the planned contract including only 20 km of rehabilitation together with routine and winter maintenance. A fourth PBC was being prepared under the ADB-funded East–West Highway Improvement Project, covering 150 km of roads. However, these PBC pilots were later canceled and removed from the projects.

Although the government is pleased with the results of the first OPRC pilot and has planned several successive pilots, the subsequent OPRC pilots have not been awarded and in some cases have not even been tendered. The government remains keen to apply the PBC approach to the government-funded maintenance contracts that continue to be based on unit rates, applied under 3-year contracts. These output-based maintenance contracts tend to have budget overruns, and the Roads Department has indicated that it is keen to move to PBCs to make the budgeting more predictable. However, application of the PBC under government procurement and financing procedures is not straightforward and would require adjustments to legislation and standards.

Performance standards. The set of performance standards prepared and used in the Kakheti pilot project are linked to payment deductions in case of noncompliance. The performance standards include requirements for reporting and for collecting traffic and inventory data each year. Deductions are applied for each 1 km section with fixed deduction percentages defined for each performance standard. For important performance standards, the deduction percentage is set at 100% (large potholes, significant edge breaks or height differences with the shoulder, missing safety measures), providing a clear incentive to comply with these standards. However, for the other performance standards, the deduction percentages are relatively low and only noncompliance with every single standard would lead to deduction of the full payment for that 1 km section. This means that the contractor can be assured of a significant portion of the monthly payment, even if no maintenance is carried out. Deduction percentages increase exponentially if they are not corrected before the next inspection. The contractor is provided with response times within which defects need to be corrected, causing the contractor to await the results of the inspection before carrying out repairs rather than acting proactively. Winter maintenance has its own set of performance standards, consisting of a combination of maximum thresholds (e.g., snow depth) and response times for snow removal and spreading of salt and sand.

Building on the lessons learned from the Kakheti pilot, the response times were removed for the Guria pilot, and deductions were to be applied immediately if a noncompliance was identified during the inspection. The deduction system has also been drastically changed in the Guria pilot. The performance standards are grouped into 10 categories related to the road condition,[19] complemented by a further six categories related to management performance (planning and reporting). For each category, a noncompliance score is calculated by multiplying the identified number of noncompliance with individual performance standards, by the weighting factor for the category concerned and a second sub-weighting factor related to delays in responding to noncompliance. The noncompliance scores for the different categories are added up to calculate the total noncompliance score. If the total noncompliance score is below a specified minimum threshold, no deduction is applied and the contractor receives the full payment. Where the total noncompliance score exceeds a specified maximum threshold, the deduction percentage is set at 100%. The minimum and maximum threshold are higher in the first months and gradually reduce as the road conditions are improved and the contractor gains experience. If the total noncompliance score is between the minimum and maximum threshold, a complicated formula is used to determine the deduction percentage.[20] This calculated deduction percentage is only applied to 80% of the monthly lump-sum payment, with 20% not subject to deductions. This system of calculating the deductions to monthly payments in case of noncompliance with the performance

---

[19] Pavement, shoulder, drainage, signage, structures, emergencies, vegetation, signs, markings, safety barriers, and other road furniture.
[20] Deduction percentage = $-0.0091X^2-0.097X+100$, whereby X is the noncompliance score above the minimum threshold.

standards is complicated, and the relationship between the level of compliance with the performance standards and the approved payment is not very clear. This is possibly one of the reasons for the high bid prices received.

Way Forward

Road asset management system. The Roads Department has developed a good working RAMS over the past decade, with data collected every year, and the RAMS fully integrated into the annual planning and budgeting procedures. The data collection and functionality of the RAMS database have been upgraded over the years as capacities and data needs increased, and the RAMS currently forms the basis for annual and multi-annual planning by the Roads Department. So far, however, the data collection has been limited to basic road inventory data, pavement condition data, and traffic data. The Roads Department is keen to expand the inventory data to comply with their standards on road passports, and to expand the condition data collected for bridges and other large structures. The Roads Department is also keen to further develop the RAMS database to allow for remote access, facilitating the editing and viewing of the available data. The software will also be upgraded to further improve the GIS functionality, allowing improved combination of GPS coordinates and linear referencing (chainage). The Roads Department is counting on development partner support to help them achieve this functionality improvement of their RAMS system. The RAMS unit currently has only three staff. The expected increase in workload related to the International Road Assessment Program assessment and the inclusion of detailed road passport inventory data will require an increase in staff numbers or a move toward regular outsourcing of certain RAMS activities.

Performance-based road maintenance contracts. The Roads Department is implementing a pilot PBC, but tendering was successful only on the second attempt. The second pilot was canceled after tendering failed, and a planned third and fourth pilot have been canceled before tendering started. Although the second pilot built on important lessons learned from the first pilot, it introduced a complex system for calculating payments and related deductions. Further improvement of the PBCs is still required, especially regarding the performance standards and their effect on deductions. This relationship needs to be simple and easily understandable to facilitate the preparation of bids and the compliance with the standards by contractors. The performance standards and related deductions also need to provide a proper balance between an acceptable road condition and the costs of achieving this. This will require a detailed review of the ongoing pilot, and further piloting of an improved and simplified system. A next step will also involve the piloting of the PBC approach in government-funded maintenance contracts. This may face additional hurdles related to existing legislation and standards that can only be identified through a thorough review and detailed preparation of a government-funded pilot.

Kazakhstan

Road network. Kazakhstan has a road network of approximately 97,000 km. Order 315 of 2015 of the former Ministry of Investment and Development formally registers 24,826 km of republican roads, including 5,378 km of international highways, 13,028 km of strategic republican roads, and 6,419 km of other republican roads. By the end of 2019, 88% of the republican roads were considered to be in good or fair condition. In terms of road activities, distinction is made between maintenance (excluding pavement and structure repairs), current repairs (pavement and structures), midterm repairs (periodic maintenance), capital repairs (rehabilitation), and (re)construction.

Institutional framework. Responsibility for the republican road network currently lies with the Ministry of Industry and Infrastructure Development (MIID) through its Committee of Roads (COR). Management of the republican road network is carried out by joint stock company Kazavtozhol that signs annual management agreements with the COR. Kazavtozhol also acts as the toll operator on four sections of tolled republican roads with a length of 682 km, where it carries out all current repair and maintenance. In non-tolled republican roads, current repairs and maintenance are carried out by Kazakhavtodor that was transformed from an enterprise under the COR into a limited liability partnership in 2017 and subsequently privatized in 2019. An initial plan to unbundle Kazakhavtodor into many smaller companies that would be privatized and compete with each other was scrapped, and instead Kazakhavtodor was privatized as a single company and transferred into a trust management under Arai Oil limited liability partnership. Kazakhavtodor currently has a 3-year grace period up to 2022, during which it will continue to be directly contracted to carry out all current repair and maintenance. Midterm and capital repairs as well as (re)construction are tendered out competitively to contractors. Road testing is carried out by the research institute KazdorNII under the MIID. In addition, there are 16 Oblzhollaboratories at regional level that have recently been merged into the newly created republican state enterprise National Center for the Quality of Road Assets (NCQRA). Among other things, the NCQRA is responsible for works and material quality control in all public roads, but also for surveying the road network and for managing the road asset management system that is under development.

Financing. (Re)construction of republican roads is financed from annual allocations from the Republican Budget, the Welfare Fund, and external loans. By law, the financing of repair and maintenance from external loans is not allowed, and these activities rely on domestic financing from the Republican Budget and the Welfare Fund. In the case of tolled roads, current repair and maintenance are financed from the toll revenues. Although expenditure on repair and maintenance doubled from T41 million in 2015 to T81 million in 2019, the budget allocation remained relatively steady at $220 million per year. This period did see a significant shift in the allocation of the funding, with midterm repair expenditure increasing by 50% at the expense of the capital repair budget. An HDM4 strategy analysis carried out in 2020 for 7,779 km of republican roads identified a budget need of T1,209 billion over the period 2021–2030 (average $320 million per year), with 80% of the funding going to capital repair. Extrapolation to the full republican road network would imply a budget requirement of approximately three times this amount. Proper review of the budget needs will require data collection to be extended to the full republican road network.

Road user charges. A Road Fund existed in Kazakhstan until 1998, which was financed from a corporate road tax (0.5% of corporate turnover), a fuel tax (T3.0 per liter), and a foreign vehicle entry fee for freight vehicles. In 2014, revenue collection from road user charges amounted to $140 million from fuel excise tax; $24 million from vehicle excise tax; $24 million from foreign vehicle entry fees; $880 million from customs duties on vehicles; and $55 million from drivers' licenses, vehicle registration, and vehicle technical inspections. Kazakhstan currently collects sufficient road user charges to allow financing of road maintenance and repair needs, but only a small portion of these are earmarked for the road sector. Instead, Kazakhstan is currently focusing on road tolling as the prime source of road user charge revenue for the road sector and currently has 682 km of roads under four toll sections. Tolls are collected by Kazavtozhol, which uses the toll revenue to carry out maintenance and current repair in the tolled road sections. The government plans to put an additional 5,700 km of recently improved roads under tolling, with a further 5,300 km to be put under tolling after completion of reconstruction works. According to Kazavtozhol's 2013–2022 Strategy as updated in 2018, tolling will ultimately cover 15,917 km of roads (over 70% of the republican road network).

Road Asset Management System

Development. Support to RAMS development is being provided under the World Bank-funded South–West Roads Project. A technical assistance consultant was contracted in 2016 to work with COR, Kazavtozhol, and KazdorNII to procure data collection equipment, carry out data collection, develop the RAMS database, and prepare multi-annual plans. The contract was extended in 2019 and is expected to run until the end of 2021. The management of the RAMS and the collection of RAMS data has now been centered in the NCQRA, which receives an annual budget allocation to cover its costs for quality control and road surveys ($5.5 million in 2020). ADB is currently funding a technical assistance (TA 6635-KAZ) to provide institutional support to the NCQRA, which will develop guidelines and draft amendments to national regulations for road maintenance, asset management, and economic analysis, including traffic forecasting.

Data collection equipment. Data collection is carried out by the Oblzhollaboratories under the NCQRA that have a total of 16 survey vehicles (1 in each Oblast + Almaty and Nursultan cities), of which some are equipped by Spetsdortechnica (Спецдортехника) and others by Rosdortech (Росдортех). The vehicles include odometers (chainage), laser profilometers (roughness), transverse lasers (rutting), inclinometers (cross slope), downward-facing linear cameras (pavement defects), front and rear panoramic cameras (inventory and defects), and GPS, as well as two trailers with FWD and for measuring skid resistance. Data from these survey vehicles can be directly uploaded to the RAMS database. Additional mobile testing vehicles were procured by KazdorNII in 2014, including a fully equipped Dynatest vehicle with FWD, ground-penetrating radar, laser profilometer, transverse lasers, cameras, GPS, and a fully automated laser crack measurement system (LCMS). A second NPO Region (НПО Регион) vehicle includes cameras, GPS, and specific software dedicated to asset recognition and positioning for road inventory data collection in support of updating the road passports. However, data from these two vehicles cannot be directly uploaded to the RAMS database and is transferred through a structured MS Excel file.

Data collection. Between 2016 and 2020, data were collected for 1,415 km of republican roads in Kostanay Oblast with ADB support as part of the preparation of a performance-based pilot project. This included road inventory and condition data as well as data for bridges and culverts. In 2018, the Oblzhollaboratories, with World Bank support, collected data for 7,779 km of republican roads (2,025 km of international roads, 4,468 km of strategic roads, and 1,286 km of other republican roads). In 2019, with the merger of the Oblzhollaboratories with the NCQRA, only limited additional data were collected. Data collection includes inventory data (geometry, GPS, sight distance, video) and pavement condition data (roughness, deflection, rutting, skid resistance, and pavement defects). The focus has been on roughness data (available for 10,770 km), with data coverage for other aspects remaining much lower (rutting 4,336 km, visibility 3,069 km, longitudinal slope 5,383 km, curvature 8,384 km). To a large extent, this is due to the fact that these other data types require post-processing of the collected data, which can be very time-consuming. Discussions are ongoing to procure equipment and software to partially automate this process. Some road passport data were collected by KazdorNII using their survey vehicle and provided in structured MS Excel files for entry into the RAMS database. Additional data are collected from other sources, including traffic and accident data. Although all data collection was financed under the project, the MIID has a specific budget line for financing quality control and network surveys that may be used to finance future replication.

Database. A tailor-made RAMS database was developed under the World Bank-funded South–West Roads Project. The initial database did not have the required functionality and a second database was developed in 2019–2020. The database allows pavement survey data to be entered from the Oblzhollaboratory survey vehicles as well as road passport data provided by KazdorNII. The database includes a diagnostics module for the preparation of standard reports regarding road network conditions and road passports in line with existing standards, as well as customizable reports. The report data can be exported in PDF or MS Excel format. The database is remotely accessible as a webpage and through a mobile app. Users are assigned different user rights for viewing and editing data and all activity is logged. The database can be viewed in Kazakh or Russian language and certain parts are available in English. A GIS module was added to the database to allow data to be presented on maps. This allows roads to be presented by technical category, pavement type, ongoing works, road restrictions, traffic level, and road facilities. In addition to the database, two mobile applications were developed. The first provides access to the RAMS database and GIS module (excluding data analysis functions) and requires user authentication, while the second provides access to simplified mapping of certain RAMS features (roads, services, pavement conditions, traffic restrictions, and ongoing works) and can be used by the wider public and does not require user authentication.

Data analysis. The data analysis includes a basic analysis module in the RAMS database. This prepares standard reports and graphs using the RAMS data including an assessment of road network changes (visual assessment of pavement, roughness, pavement strength, and rutting); road safety levels; and road traffic capacity. These can be prepared for the entire network and for specific roads. Actual condition data are compared against condition categories as defined in national standards, or against specific thresholds set by the user. The RAMS database also includes a cost module that uses unit costs to prepare preliminary cost estimates. This is linked to a simple decision matrix that allows different treatments to be defined based on the severity and type of defects (roughness, rutting, deflection, surface defects). Based on the selected defect, thresholds, and treatments, preliminary budget needs can be determined for the entire network or for a subnetwork, differentiated by reconstruction, capital repair, midterm repair, current repair, and maintenance. The results can be shown for the entire network, by road or by road section. As such, the integrated planning module only serves to give an indication of the types of treatment required, but does not serve to prepare actual plans or budget requests. The use of the RAMS for planning also contradicts existing standards that prescribe certain treatments to be carried out after a specific number of years.

Further analysis for planning purposes is carried out by organizing the data into 1 km sections and exporting the data to an MS Excel file with 112 data fields. This is validated and organized into homogeneous sections based on roughness, rut depth, and traffic, after which it is imported into HDM4 for analysis. An HDM4 strategy analysis was carried out using the 2018 data, which formed the basis for the preparation of a 3-year plan. However, this analysis was limited to the 7,779 km for which data were available and the results are not yet considered suitable for preparing a plan for the entire road network. Data collection for the full republican road network followed by an HDM4 strategy analysis and program analysis is required to prepare a network plan and to determine future financing needs. The lack of a Russian language interface for HDM4 is complicating its use by the NCQRA.

Performance-Based Road Maintenance Contracts

Pilots. Kazakhstan does not yet have experience with performance-based contracting for road maintenance. ADB has been supporting the preparation of the Road Maintenance Sustainability Project that is to include a large PBC component. The PBC component was supposed to cover all 1,415 km of republican roads in Kostanay Oblast. This would include four PBCs, of which one would be combined with the upgrading of 161 km of the M36. The performance-based maintenance was to be for a period of 8 years. The full coverage of the maintenance needs for all republican roads would require an increase in annual maintenance expenditure from $22 million to $30 million per year for Kostanay Oblast. However, the project is still pending approval as legislative provisions prohibit borrowing money for maintenance, and government funding to maintenance contracts cannot be committed for such an extended period. Kazavtozhol reportedly introduced a pilot contract with performance-based elements on a 42 km road section in Aktobe Oblast, although little information is available regarding this pilot. There is also an operation and maintenance obligation in the contract for the construction of Big Almaty Ring Road (Большая Алматинская кольцевая автомобильная дорога [BAKAD]), although this involves a concession contract that falls under the Public–Private Partnership (PPP) Law. The PPP Law appears to provide more opportunities for implementing performance-based maintenance, as it is not as strictly regulated with respect to norms for road maintenance spending. Although PPPs offer an opportunity to introduce PBCs, they also introduce other requirements that complicate their application compared to traditional PBCs.

Performance standards. As part of the preparation of the Road Maintenance Sustainability Project, draft PBC and bidding documents were prepared. However, these bidding documents still require improvement. The performance standards include response times, which may cause contractors to wait for the inspection results rather than carrying out the maintenance activities proactively. Routine, winter, and periodic maintenance are combined into a single performance-based, lump-sum payment with deductions applied to this combined payment in case of noncompliance with any of the performance standards. This introduces a high risk for contractors inexperienced with PBCs and is likely to lead to high bid prices and increase the danger of contractor default. Some of the performance standards need adjustment to properly reflect the desired road conditions (e.g., maximum size of potholes currently set at 0.25 square meter [m^2], equivalent to a diameter of 60 cm). The deduction percentages are very low and are not in line with the seriousness of the defect or the cost of correcting it. These deduction percentages need to be increased to ensure that there is sufficient financial incentive for contractors to carry out the maintenance activities in a timely manner. A review of the performance standards should be carried out prior to applying them in a new PBC pilot.

Way Forward

Road asset management system. The development of the RAMS in Kazakhstan is well underway. A RAMS database has been developed that is linked to existing standards regarding spring and autumn surveys, road diagnostics, and road passports. Data collection equipment has been procured and a RAMS unit has been created under the NCQRA. However, challenges still remain. Although data has been collected for a considerable portion of the republican road network, this only covers some of the data needs. Considerable support is still required to complete the inventory and condition assessment for the full republican road network. This will benefit from additional equipment and software to automate part of the defect and inventory recognition. The second important area for future support is the data analysis and the use of the analysis results for preparing annual and multi-annual plans for the republican road network. This requires extensive training in the use of HDM4 as well as the definition of procedures for incorporating analysis results into the planning process. The current technical assistance from the ADB is expected to address some of these issues. There is currently a very strong basis for a sustainable RAMS, but without this additional support there is a significant risk that it may fail or that it will not affect the planning and budgeting process.

Performance-based road maintenance contracts. Kazakhstan is going through a transition regarding the current repair and maintenance of republican roads. Kazakhavtodor has been privatized and after 2022, all current repair and maintenance contracts will be tendered competitively. Kazavtozhol is carrying out current repair and maintenance on tolled roads, but as the tolled road network increases in size this will likely also start to be contracted out. Thus, there is an urgent need of introducing contracting modalities that are well suited to repair and maintenance of republican roads and involvement of private sector contractors. PBCs have internationally proven to have important benefits, allowing contractors to improve their efficiency through proper planning and facilitating budgeting and supervision by the contracting agency. Given the changing context of current repair and maintenance implementation, now is a good time to pilot performance-based contracting. According to Kazavtozhol, the new repair and maintenance contracts to be tendered will apply a "defect-free" concept following performance-based principles. This is an important change and one which should receive appropriate support from development partners to ensure that the performance-based aspects are properly designed and fit for purpose. For PBCs to be possible, however, the issues of government lending and multi-annual budgeting for maintenance will need to be addressed. The draft bidding documents for PBCs should also be reviewed in detail to ensure they result in a proper balance of responsibilities and risks and lead to acceptable bid prices.

Kyrgyz Republic

Road network. The Kyrgyz Republic has a road network of approximately 35,000 km, of which 18,993 km are considered to be public roads. These public roads include 4,216 km of international roads, 5,673 km of national roads, and 9,104 km of local roads. Based on the data collection for paved roads carried out in 2018–2019, approximately 60% of international roads were found to be in good condition, compared to 20% of national roads and just over 10% of local roads. In terms of road activities, distinction is made between maintenance (excluding pavement and structure repairs), current repair (pavement and structures), midterm repair (periodic maintenance), capital repair (rehabilitation), and (re)construction.

Institutional framework. Up until recently, the Ministry of Transport and Roads (MOTR) was responsible for the management of the public road network (international, national, and local roads). In 2021, the ministry was transformed into the Ministry of Transport and Communications (MOTC). Under the MOTC, the Road Administration Department is in charge of policies, norms, and regulations for the road sector. The Road Maintenance Department (DDH)[21] is set up as a separate legal entity under MOTC and is responsible for management of the road network, supported by four oblast-based regional offices,[22] four corridor-based UADs,[23] and one state directorate responsible for the main Bishkek–Osh road, all set up as independent legal entities. Implementation of road maintenance and repair is done by the 57 local road maintenance units (DEU)[24] that are set up as separate legal entities, as well as two state enterprises under the Bishkek–Osh State Directorate. Works beyond the capacity of the DEUs are contracted out. Planning is traditionally done by the DEUs based on visual assessments of road conditions carried out during the spring and autumn surveys, with plans reviewed and approved by the regional offices/UADs and ultimately by DDH. With the development of the RAMS, the planning is expected to become more centralized, based on objective data measurements. As such, a more important role is expected to be played by the Asset Management Section under DDH. The former MOTR Technical Training Center was transformed into the self-supporting state institute Production Innovation Centre (PIC) in 2017, and is currently involved in the RAMS data collection and database management on a contract basis with MOTC.

Financing. The trunk road network in the Kyrgyz Republic suffers from a serious maintenance backlog, especially in the national and local road networks. According to an HDM4 strategy analysis carried out by ADB in 2019 based on data collected for most of the paved road network, the optimal maintenance strategy would require a budget of $220 million per year during the first 5 years to address backlog maintenance, followed by $70 million per year thereafter. This is in line with a similar assessment carried out by the European Bank for Reconstruction and Development (EBRD) in 2014. The actual annual road maintenance and rehabilitation budget was only $28 million in 2018, equivalent to only 40% of the optimal annual maintenance budget (25% if the budget for backlog maintenance is included). Of the available maintenance and rehabilitation budget, approximately 40% is spent on rehabilitation and previous obligations, with the rest spent on current repairs, routine maintenance, and winter maintenance, as well as maintenance of the DEUs themselves and other services. Expenditure on midterm repairs is extremely limited. Although a Road Fund Law exists

---

[21] Департамент Дорожного Хозяйства (ДДХ).

[22] Производственно-Линейное Управление Автомобильных Дорог (ПЛУАД): Chui, Naryn, Issyk-Kul, and Talas (formerly Production and Linear Management of Roads or PLAUDs).

[23] Управления Автомобильных Дорог (УАД): Bishkek–Naryn–Torugart, Osh–Batken–Isfana, Osh–Sarytash–Irkeshtam, and Jalal Abad–Balykchy—these also take care of surrounding roads.

[24] Дорожно-Эксплуатационное Учреждение (ДЭУ).

with earmarked road user charges, this is not applied in practice and the road maintenance budget depends on annual allocations from the general budget. A study on road user charges carried out by EBRD in 2014 found that actual budget allocations to road maintenance exceed the revenue from the road user charges earmarked to the Road Fund. However, with maintenance and rehabilitation budgets remaining the same or even decreasing since 2014 and revenues from these road user charges increasing over the same period, this is likely no longer the case.

Road user charges. The Road Fund Law was issued in 1998 and amended in 2008. This earmarks 50% of fuel excise taxes, 90% of vehicle registration fees, 100% of overweight and oversized vehicle fees, and 100% of toll revenues for maintenance and development of the road network, in addition to revenue from the general budget and from development partners. Previous earmarked revenue from a vehicle tax and from a road tax based on corporate turnover were removed from the Road Fund in 2008 when these taxes were replaced by a property tax. An EBRD study found the fuel excise tax revenue amounted to just over $18 million in 2012. Fuel consumption has increased since then, and more importantly, the excise tax rates have increased fivefold to Som10,000 per ton for petrol ($0.08/liter) and tenfold to Som2,000 per ton for diesel ($0.02/liter). Current fuel excise tax revenue is therefore expected to be in the order of $100 million per year. Vehicle registration fees used to form another important revenue source, amounting to approximately $13 million in 2012 according to the EBRD study, but these fees were abolished in 2017. The EBRD study found that revenue from overweight/oversized vehicles and tolling was relatively low ($0.6 and $1.0 million, respectively, in 2012). The property tax for vehicles, although not earmarked to the Road Fund, is also an important source of revenue with $12 million in revenue in 2012. Even more important is the customs duty on vehicles, amounting to $89 million in 2012. In 2018, the Road Fund was abolished as it contradicted the budget law of that year. Discussions are now ongoing to reinstate the Road Fund under a new law. The road user charge revenue formally earmarked for the Road Fund is likely to be sufficient to cover the optimal maintenance budget of $70 million per year. This could be expanded to include revenues from other road user charges, allowing improvement works to also be partially covered.

Road Asset Management System

Development. RAMS development has been supported by different development partners over the last decade. The Japan International Cooperation Agency (JICA) supported the collection of road roughness data in the Project for Building Capacity in Road Maintenance in 2010. The World Bank's National Road Rehabilitation Project supported data collection and the development of a simple RAMS between 2010 and 2013, while the ADB-funded CAREC Corridors 1 and 3 Connector Road Project—Additional Financing is currently supporting the further development of a comprehensive RAMS system and training of DDH staff. The RAMS data collection and database management are currently being carried out by PIC under a contract with DDH. The legal basis for the involvement of PIC is lacking, however, and a shortage of contracts as a result of COVID-19 has caused PIC to lay off some of their staff, putting the capacity building and the sustainability of the data collection and database management at risk. Data analysis and planning based on the RAMS continues to be the responsibility of DDH and its Asset Management Section, although capacities are still limited.

Data collection equipment. JICA introduced the Vehicle Intelligent Monitoring System (VIMS) equipment to the Kyrgyz Republic. This consists of a GPS receiver and an accelerometer located over the vehicle wheel hub connected to a laptop computer to record GPS coordinates, IRI, and speed. This equipment was subsequently used to collect roughness (IRI) data for international and national roads under the National Road Rehabilitation Project of the World Bank. The MOTR later purchased a Rosdortech Trassa vehicle that includes a GPS receiver, odometer, laser profilometers for measuring roughness, transverse profiler for measuring rutting, and forward- and backward-looking video cameras. Under the CAREC Corridors 1 and 3 Connector Road Project—Additional Financing, the RAMS consultant used this vehicle to collect data, complementing it with their own high-accuracy GPS receiver and high-definition video camera to improve data quality. The export of the collected data from the Trassa vehicle to the RAMS database cannot be carried out automatically and requires several manual data entry steps that increase the risk of errors. The equipment on the Trassa vehicle is too sensitive for collecting data in poor condition or unpaved roads and the purchase of additional basic data collection equipment is being considered (e.g., basic GPS, camera, and bump integrator). Under the CAREC Corridors 1 and 3 Connector Road Project—Additional Financing, five DataCollect SDR radar traffic counters were also procured and used to collect traffic data. Collected traffic count data are downloaded to a smartphone application using Bluetooth, and subsequently processed to provide traffic counts for both directions indicating the length and speed of each vehicle.

The video data collected with the Trassa vehicle (with frames every 10 meters) are used for post-processing using Destia Devli software to identify inventory data (terrain, land use, pavement width, median width, surface type, number of lanes, shoulder type, shoulder width, and location reference points) using basic categories and linking this to GPS data of the video frames. The post-processing also allows surface distress data to be recorded from the video footage by projecting the backward-looking video image as a downward-facing image, and manually indicating the affected area on the video frames using the software (number and area of potholes, area of cracking, edge break, and patching).

Data collection. Under the National Road Rehabilitation Project of the World Bank, basic inventory data were obtained from existing records for approximately 18,000 km of roads, almost the entire public road network. Roughness data were collected for a total of 1,377 km of international roads through drive-over surveys using the VIMS equipment. This was combined with visual condition assessments for the same sections, identifying any surface distress. One-day traffic counts were also carried out. The data collection faced challenges with the DDH not considering it part of its tasks, focusing instead on the implementation of maintenance and repair works. Providing staff to collect the data over extended periods of time and covering the costs of fuel and per diems was also a major challenge, resulting in a reduction of the surveyed length from the originally planned 6,000 km. The project concluded that a specific decree stipulating the data collection responsibilities was necessary, together with the establishment of a dedicated RAMS unit.

Under the CAREC Corridors 1 and 3 Connector Road Project—Additional Financing, data collection is carried out by the PIC under a contract with DDH, with support provided by the project consultants. Data for 5,835 km of paved roads (one-third of the network) was collected through drive-over surveys using the Trassa vehicle in 2018–2019 (2,924 km of international roads, 1,786 km of national roads, and 1,125 km of paved local roads). Data collection included GPS, roughness, rutting, and video. All data were checked and validated before entering it into the RAMS database. This led to approximately 1,700 km of road being resurveyed due to problems with some of the data. This exercise also identified a number of discrepancies with the naming, numbering, and formal lengths of the roads. Manual post-processing of the video data involved a considerable input in terms of person days, but provided basic inventory and pavement distress data for the full 5,835 km. Traffic data were also collected for the same roads, involving 1-day traffic counts in 231 locations using the portable SDR radar equipment. Bridge data for international and national roads was transferred from the database prepared by JICA in 2014 (including inventory and condition data as well as photographs). Tunnel data were copied from the existing tunnel passport data.

Additional data collection is currently ongoing for the remaining 2,572 km of paved roads (580 km of international roads, 620 km of national roads, and 1,372 km of local roads), complemented by traffic counts in these roads at 150 locations. Further data collection has been proposed for the 10,586 km of unpaved roads (712 km of international roads, 3,267 km of national roads, and 6,607 km of local roads), but this has yet to be approved and financing obtained. For the local roads, basic survey equipment would be used; traffic counts for low-volume roads would be based on moving traffic counts using video data, rather than separate traffic counts. The additional data collection is also proposed to include updating the bridge and tunnel data and collecting data on culverts. Such structure data cannot be collected through drive-over surveys and requires stopping at each structure. Data collection is further proposed to include pavement strength testing using an FWD to aid in determining the need for rehabilitation, requiring the procurement of FWD equipment and separate FWD surveys.

Database. Under the National Road Rehabilitation Project, a simple database was developed using MS Excel, since this software was already widely used by DDH, avoided the need for specific training, and allowed a Russian language interface to be developed. Basic inventory data were entered from existing data in DDH.[25] The data were organized into the road sections that were applied by DDH based on the allocation of responsibility to different DEUs. Data on the number and length of bridges and culverts was also entered for each road section, as was data from manual traffic counts that were carried out in some roads. The road sections were organized into homogeneous segments of up to 1 km and collected GPS data had to be manually processed to fit these road segments before being imported into the database (together with any references to photographs of structures or other features). Roughness data collected using the VIMS was also entered for these 1 km segments, using a specific macro to facilitate the import from the standardized VIMS data files. The MS Excel database allowed data to be exported to Google Earth to visualize the roughness and chainage data. The database also included a number of standard reports to present the main features of the road network (length by class, technical category, surface type, PLUAD, traffic volume, roughness, surface distress, etc.) in tables and graphs. Around the same time, a Filemaker database was developed with JICA support for managing the bridge data. This included basic inventory and condition data for each bridge, as well as up to three photographs.

Under the CAREC Corridors 1 and 3 Connector Road Project—Additional Financing, a web-based RAMS database has been developed. This was initially set up as an SQL server and later migrated to an open-source spatial database with PostgreSQL and PostGIS. The formal road registry was used as the basis for the road list, introducing specific road coding as unique identifiers in the database. Data from the Trassa drive-over surveys were incorporated together with the basic inventory and surface distress data acquired from the video post-processing. Traffic data were entered as raw data by vehicle length (classified by 10 cm intervals), allowing vehicle classification to be adjusted in the database together with the conversion factors for estimating the average annual daily traffic based on the 1-day traffic counts. Bridge data from the JICA Filemaker database was also transferred. The RAMS database allows data to be viewed in the form of tables, graphs, maps, and photographs. The database further allows selected data to be exported as comma separated values, and allows standard reports and standard maps to be prepared to display road asset inventory, road network condition, and works programs. Differentiated user access levels were introduced in the database, with DEUs and regional offices/UADs only able to access the data for the roads under their responsibility. For each user, the rights to edit and change the data can also be defined. The interface is available in English, Russian, and Kyrgyz languages. The PIC is responsible for entering the collected data in the database. DDH staff are currently receiving training from the project consultants in the use of the database.

Data analysis. The MS Excel database developed by the National Road Rehabilitation Project included a basic costing and prioritization tool. A treatment decision matrix defined the optimal treatments for each road section based on traffic volume and road condition (IRI, rutting, cracking, and potholes). Treatments were only defined for bituminous pavements and varied from patching and crack sealing to chip seals, asphalt concrete

---

[25] Road number, section number, name, region, PLUAD, DEU, technical category, length by surface type (recorded and according to GPS), width, number of lanes, year of construction, and last treatment.

overlays, and full reconstruction. The proposed optimal treatments were determined through an economic analysis at network level (strategy analysis) using RONET. The cost of the selected treatments and resulting works program could be estimated using unit rates for each treatment entered into the database, allowing a budget estimate to be developed. The prioritization of the proposed road sections and related treatments was based on a Condition Index defined by the roughness and the degree of rutting, cracking, potholes, edge breaks, and deformation, which was complemented by a Functional Index that defined the importance of the road based on the road class, technical category, traffic volume, and surface type. Together these two indices formed the Priority Index that was used to prioritize the road sections for budget allocations. Data from the database could be exported for further analysis using RONET, with the database automatically providing the data in the required format. The MS Excel database provided standard output reports on the volume of works required, organized by treatment type and by PLUAD/UAD. Although the database was comprehensive, the final evaluation of the project found that it was not being used by DDH to inform the planning and decision-making process.

The system currently being developed under the CAREC Corridors 1 and 3 Connector Road Project—Additional Financing builds on this concept with the aim of developing a decision-support tool for preparing a prioritized works program. The main changes are in the database itself and the presentation of the data it contains, with the planning module remaining largely the same. The treatment decision matrix has been copied with only minor adjustments following an HDM4 strategy analysis. The priority index continues to be used as the basis for planning, with some adjustments to the criteria used for calculating the underlying condition index (roughness, rutting, cracking, potholes, patching, and edge breaks), and functional index (road class, technical category, traffic volume, population density, black spots, and social importance). Cost estimates continue to be based on unit rates, but can be adjusted for individual roads to take account of local differences. The planning tool makes use of a decision matrix to determine the periodic maintenance and rehabilitation work program based on collected data, which is complemented by a separate routine maintenance and repair program that is prepared by the DEUs. The planning methodology has only been prepared for paved roads, and a methodology for unpaved roads still needs to be developed. The project consultants are assisting DDH in using the planning module to prepare annual and 5-year rolling work programs. As opposed to the traditional approach where DEUs prepare the annual work programs based on visual assessments and submit these to the regional offices/UADs and subsequently DDH for approval, the annual work programs are now prepared at the central level and shared with the regional offices/UADs and DEUs for verification and comments. After initial review by DDH, additional features were added to respond to their needs, including the prevention of unauthorized changes to the works program once approved, and the presentation of the work programs in the format of the official Title List that forms the basis for the annual budget request.

An important achievement in the Kyrgyz Republic is the issuance by MOTC of an Amendment to the Order of the Ministry of Transport and Roads of the Kyrgyz Republic on Approval of the Regulations and the Procedures for Planning and Carrying Out Repair Works and Maintenance of Public Highways and Road Structures in the Kyrgyz Republic in June 2021. This new order amends the previous 2017 order and introduces the mandatory use of the RAMS in planning the repair and maintenance of public highways and road structures, and applies amendments to the classification of the types of work performed during reconstruction, capital repair, and midterm repair. The Law on Highways is also currently under revision, and one of the proposed amendments is to stipulate that planning of works for the repair and maintenance of public roads must be carried out with the mandatory use of the RAMS. This will provide a very good legal basis for the continued use and operation of the RAMS.

Performance-Based Road Maintenance Contracts

Pilots. Under the ADB Regional Road Corridor Improvement Project (2008–2014) that covered both the Kyrgyz Republic and Tajikistan, five PBCs were to be piloted in each country, covering 500 km of roads with contract periods of up to 4 years. The contracts were to be tendered to domestic private sector contractors with 100% financing from the government budget and technical assistance provided under ADB financing. After significant delays, two PBCs were awarded in Tajikistan in 2013, but in the Kyrgyz Republic the procurement faced legal and financial difficulties with the government unable to secure the required budget for the 4-year period, and it was decided to transfer the piloting of PBCs to the CAREC Corridor 3 Bishkek–Osh Road Improvement Project.

Around the same time, the Central Asia Road Links Program of the World Bank also aimed to pilot performance-based contracting. Due to the problems in piloting PBC pilots with private sector involvement, the program instead piloted a service-level agreement (SLA) between DDH and the Osh–Batken–Isfana UAD covering the routine summer, winter, and emergency maintenance of the entire road corridor.[26] The aim was to move away from the traditional practice of payment according to the inputs used, toward a payment according to the resulting performance of the road. The work of the UAD was financed from the Republican Budget (covering staff, equipment operation, and material purchase costs), with road maintenance equipment provided under World Bank financing. The SLA was unclear on roles and responsibilities of the UAD, the amount of funding provided to the UAD was insufficient to comply with the performance standards, and the road maintenance equipment was only provided at the end of the agreement. As a result, the SLA was only operational for 1 year from April 2014 to March 2015 and was not renewed. The approach did lead to a significant increase in maintenance funding for the corridor concerned, although this remained below what was required.

Under the ADB CAREC Corridor 3 Bishkek–Osh Road Improvement Project, a 3-year PBC was piloted in a 68.5 km section of the Bishkek–Osh road from Karabalta to Sussamyr (61–129 km). The pilot was funded from a combination of 50% government budget and 50% ADB grant financing. After the first tender failed, the second tender was successful and resulted in a contract award to a domestic private sector contractor for $4.3 million (January 2018–December 2020). The contract sum included initial rehabilitation of 8 km and periodic maintenance of 9 km (74% of the contract sum), as well as emergency maintenance and dayworks (3%) that were paid separately on a volume basis. Current repair of the pavement (9% of contract sum) involving pothole patching, crack sealing, and large landslides were also paid on a volume basis. Snow removal (6% of the contract sum) was paid based on the cleared road length and the number of days this was required. The only activities contracted on a full performance basis included management activities (traffic safety, reporting), routine maintenance of the road and structures (temporary pothole patching with aggregate, maintenance of signs, clearing of culverts and drainage ditches, small concrete repairs, vegetation control, slope stabilization, removal of obstacles) and basic winter maintenance activities (placing snow poles, cleaning signs, patrolling, traffic management). The performance-based component formed only 7% of the contract amount, with an average cost of $1,500/km/year.

In 2020, two additional PBCs were awarded under the CAREC Corridors 1 and 3 Connector Road Project—Additional Financing. These cover the road section from Balykchy to Kochkor (0–43 km) and from Kochkor to Epkin (62–89 km). The contracts include the full rehabilitation of the road sections over a period of 2 years, followed by 5 years of performance-based maintenance and repair. The two contracts were awarded to a Chinese contractor for a total amount of $40.2 million, of which the performance-based maintenance and

---

[26] In order to cover the full corridor, two DEUs were transferred from the Osh–Sarytash–Irkeshtam UAD.

repair made up only 7.5%, equivalent to an average cost of $8,700/km/year. Contrary to the previous PBC pilot, all works except the initial rehabilitation are paid on a performance basis (a 10% provisional sum for dayworks and emergency works also involves payments on a volume basis). The performance-based activities include winter maintenance, routine maintenance, and current repairs to the pavement and structures. Even periodic maintenance (seals and thin overlays) is included in the performance-based lump sum, to be applied where the roughness of the newly rehabilitated pavement exceeds a certain threshold. As such, the scope of the performance-based activities is far greater than under the previous PBC.

Performance standards. The performance standards in the service-level agreement (SLA) with the UAD were not well defined and lacked penalty measures in case of poor performance. Actual compliance with the performance standards was low, reaching only 91.4% at the end of the agreement.

Under the CAREC Corridor 3 Bishkek–Osh Road Improvement Project, the performance standards differentiated between standards for contract management, pavement, signage, drainage, vegetation control, structures, slopes, and winter maintenance. Different performance standards were applied in summer (1 April–31 October) and winter (1 November–31 March). Not all performance standards were properly defined. For instance, potholes were not allowed to exceed 0.5 m^2 in size, which is equivalent to a pothole diameter of 80 cm and is clearly not a relevant standard. Drainage had to be clean and free of obstacles, without a clear definition of how this would be measured. Inspections were carried out every month during summer, and after every winter event during winter season. The performance standards included response times within which the contractor was required to correct any identified noncompliance with the performance standards. This was generally set at 28 days. Two different service levels were defined for the pavement, each with its own response times, with the lower service level requiring correction of the defects before the start of the winter season (1 November) rather than within a specific timeframe. Penalties were applied as deductions to the monthly payment, with deductions applied separately for contract management, routine road maintenance, routine structure maintenance, and winter maintenance (each of these had its own lump sum). For some performance standards, the deduction percentages were applied to the relevant monthly lump-sum payment per kilometer (lump sum divided by road length), multiplied by the number of 1 km segments that were found to be noncompliant. For other standards, the deduction was applied to the full lump sum for each occurrence. This resulted in limited transparency regarding the impact of specific noncompliance, with certain performance standards having a much greater impact than others. Since routine pavement repairs were paid on a volume basis, these did not result in penalties if not carried out (uncompleted works were simply not paid). The contract did not have any major issues, with the contractor complying with the performance standards (including for winter maintenance) and only a limited number of penalties applied.

Based on this experience, ADB supported the development of PBC regulations to support the replication of PBCs under government funding. The PBC regulations were issued by Ministerial Order in 2018 and include a new set of performance standards. The regulations distinguish five different service levels, whereby the service level to be applied is to be defined in the contract. Thresholds for performance standards may differ by service level (e.g., maximum area affected by potholes ranging from 0.1 m^2 to 2.1 m^2 per 100 m of road), or may be the same for all service levels (e.g., ponding water exceeding 10 cm in depth or 2 m^2 in area is not allowed for any service level). The performance standards correct some of the identified issues (e.g., the maximum size of potholes is set at 0.09 m^2, equivalent to a pothole diameter of 30 cm), but continue to include issues related to measurability and achievability (e.g., culvert sedimentation is not allowed). The PBC regulations allow for inspections of the full road length or a sample road length. Contractors are given a response time to correct any identified defects, after which the road is again inspected. Deductions are applied only if the contractor fails to correct identified defects within the allocated response time. The response times and deductions are not defined in the PBC regulations.

The two PBCs awarded under the CAREC Corridors 1 and 3 Connector Road Project—Additional Financing do not make use of the performance standards defined in the PBC regulations, and instead introduce yet another set of performance standards. These distinguish between pavement, shoulders, drainage, roadside, safety, structures, and winter maintenance. Further improvements are introduced to the standards (e.g., the maximum pothole size is set at 20 cm diameter). However, several standards still do not have measurable indicators and thresholds defined, complicating the inspection. There is one single lump-sum payment for all performance-based activities, and payment deductions are calculated by multiplying the deduction percentage for a specific performance standard by the monthly lump sum per kilometer and by the number of 1 km road segments that are found to be noncompliant. A set of five important performance standards related to potholes, drop-offs, safety signs, and guardrails have deduction percentages of 100% for the 1 km road segments where the thresholds are exceeded. The deduction percentages for the other performance standards remain very low and do not provide a proper incentive for the contractor to comply with the standards. For routine maintenance and repair, the deduction percentages add up to 100%, implying that the contractor would need to fail compliance with all 15 performance standards in a specific 1 km segment before the full payment is deducted. With deduction percentages varying from 4%–12%, this means that noncompliance in any 1 km road segment would lead to a deduction to the monthly payment of only $30–$90. The deduction percentages for winter maintenance also amount to 100%,[27] of which half are related to preparation for the winter season and regular patrolling. Although the deduction percentages are low, failure to correct the defects will lead to a 200% increase in the applied deductions each month, ensuring that contractors will ultimately comply even if the initial deduction is limited. In addition to the deduction percentages linked to these performance standards, a separate deduction calculation is applied if the actual roughness exceeds the maximum threshold (based on roughness measurements carried out each month). The contractor is further required to prepare plans and reports, but compliance is not linked to payment deductions. The contractor is provided a response time to correct any noncompliance identified during the inspection, with stipulated response times ranging from 1–90 days. Only if the contractor fails to correct the defect within the response time, does the deduction become permanent and unrecoverable. The exception are the five important performance standards with 100% deduction rates, where permanent deductions are applied immediately. The compliance is inspected during monthly formal inspections.

Way Forward

Road asset management system. The RAMS database and planning modules are still being finalized and will require extensive testing. Data has been collected for only one-third of the network and is ongoing for the remaining 2,571 km of paved roads. Further data collection is being considered for the 10,586 km of unpaved roads. Data collection for the unpaved roads involves a considerable exercise and will require adjustment of the types of data to be collected and procurement of suitable equipment. Regular updating of the condition data will require annual data collection, and annual budgets should be allocated for this purpose, complemented by standardized procedures for involving PIC in the data collection. This data collection should also include updating of the data for bridges and tunnels. Data for culverts may also be collected, but is not considered a high priority. Only through regular data collection can the sustainability of the RAMS be ensured and the capacities within PIC and DDH be maintained. The inclusion of unpaved roads will also require the expansion of the planning module for such roads. The calculation of the condition index and the related treatment decision matrix will need to be amended for use in unpaved roads, with less focus on roughness and more on surface defects. The main challenge in the further development of the RAMS, however, is likely to its institutionalization. DDH and PIC have been involved throughout the development and operation of the

---

[27] These are applied on top of any routine maintenance and current repair standards that need to be complied with during the winter season.

RAMS, but the use of the RAMS as the basis for annual work programs and budgeting has yet to be tested, and continued support will be required to further develop the capacities to operate it, to ensure that annual budget allocations are made for financing its operation, and to properly integrate it into existing planning and budgeting procedures. The recent order issued by MOTC and the proposed amendment of the Law on Highways are important steps in the institutionalization of the RAMS. The RAMS analysis will also show a clear need for additional funding, which may be addressed through the reestablishment of the Road Fund and the earmarking of suitable road user charge revenues. The RAMS is likely to lead to a shift in funding allocation toward midterm repairs (periodic maintenance), which will require suitable capacities to be developed in the contracting industry to respond to the increased demand.

Performance-based road maintenance contracts. PBCs have been piloted with maintenance units and private sector contractors in the Kyrgyz Republic, with overall positive results. Various sets of performance standards have been applied, and issues with the measurability, achievability, and relevance of the standards continue to exist. The low deduction percentages and the use of response times to correct identified noncompliance mean that there is limited incentive for the contractors to carry out works proactively, preferring instead to await the results of the inspections and correct only those defects that have been identified. The consistency of the different performance standards, the methods for their inspection, and the application of deductions in case of noncompliance requires further improvement, ensuring a coherent and transparent system is developed that is easily understood and applied. Further attention should be given to PBCs covering winter maintenance, routine maintenance, and current repair, with only limited rehabilitation or periodic maintenance works. Roads in good and fair condition fall into this category and should be prioritized for performance-based maintenance. As this will likely attract domestic contractors rather than international contractors, this should be accompanied by a program of capacity building.

Mongolia

Road network. Mongolia has a road network of 48,538 km, including 12,710 km of international and state roads. The majority of these international and state roads are unpaved, and over half the network consists of basic earthen tracks. The remaining 35,828 km are local roads that are almost all earthen tracks. The Law on Roads distinguishes between maintenance (excluding pavement and structure repairs), routine repair, regular repair (periodic maintenance), major repair (rehabilitation), emergency repair, and (re)construction.

Institutional framework. The international and state roads are managed by the Ministry of Roads and Transport Development (MRTD), through the state-owned enterprise Road and Transport Development Center (RTDC). Since 2010, major repair and (re)construction are contracted out to licensed legal entities from the private sector through competitive bidding and supervised by independent consultants. Maintenance as well as routine, regular, and emergency repairs are managed through the RTDC's Road Maintenance Department, with its six regional supervision engineers and two mobile patrol officers. Implementation of road maintenance, routine, regular, and emergency repair is by the 20 state road maintenance companies (AZZAs), although this may be contracted out to the private sector where the capacity of the AZZAs is insufficient. The RTDC's Toll Collection, Road Use and Traffic Control Department is responsible for managing the RAMS for the international and state road networks.

Financing. The ADB report on road sector development in Mongolia estimated that approximately $25 million per year was required to finance the maintenance and repair of the international and state road networks ($11 million for routine maintenance and repair and $14 million for regular repair),[28] as well as $20 million per year for rehabilitation over a 5-year period to address the existing maintenance backlog. Road maintenance financing in Mongolia has remained very low, actually decreasing in US dollar terms from $11.6 million in 2012 to $9.4 million in 2017.[29] This is equivalent to less than 0.1% of gross domestic product (GDP), compared to a minimum international benchmark of 0.4% of GDP. With increasing budgets for road construction and upgrading, the maintenance budget also makes up an ever-decreasing portion of the total road sector budget, forming only 2% of the total sector budget in 2013. As a result of the increased investments in road construction and upgrading, the paved road network has doubled in length, resulting in a significant increase in maintenance funding needs for the paved roads. Budget requests for road maintenance are based on budget estimates prepared by the road maintenance companies. Annual budget allocations tend to cover only 15% of the budget required. Budget allocations are used mainly for routine maintenance and repairs, while funding for regular repair tends to be very limited. Proposed regular repair plans prepared for 2013 and 2014 under the ADB Road Sector Capacity Development Project were not approved or funded by the government. This was reportedly due to a lack of available budget, while for the same years the budget allocation for maintenance and routine repair was doubled. In the subsequent period 2015–2017, regular repair was only carried out in total 18 km of road. According to the road asset maintenance appraisal prepared by ADB in 2018, attention within maintenance and routine repair is often focused on maintenance of shoulders and vegetation control, rather than crack sealing and pothole patching that would have a greater impact on slowing down the deterioration of the road.[30]

---

[28] This included an average of $4,200/km/year for routine and periodic maintenance of paved and gravel roads, and an allocation of only $650/km/year for the earthen tracks that make up the majority of international and state roads.

[29] The maintenance budget increased from MNT15 billion in 2012 to MNT22 billion in 2017, but due to changes in the exchange rate this was equivalent to a slight reduction in US dollar terms.

[30] ADB. 2018. Regional Road Development and Maintenance Project: Road Asset Maintenance Appraisal. https://www.adb.org/sites/default/files/linked-documents/48186-005-sd-03.pdf.

Road user charges. Up to 2017, maintenance and repair of international and state roads was financed from a Road Fund that received funding primarily from fuel tax. However, over the years, the revenue from the fuel tax decreased in real terms as a result of inflation, a fixed fuel tax rate that had not changed since 1995, and an ever-decreasing percentage of fuel tax revenue being allocated to the Road Fund.[31] The 2017 revised Road Law introduced a new State Road Fund as well as a Capital City Road Fund and a Local Road Fund. The new State Road Fund is financed from 20% of the excise tax on imported vehicles, tolls on international and state roads, a percentage of the fuel tax,[32] traffic fines applied in international and state roads, transit fees for international vehicles, fees for the use of the right-of-way, foreign loans and grants, and state budget allocations. Revenue from the excise tax on imported vehicles amounted to $84 million in 2012, but reduced to only $14 million in 2016. Tolls in international and state roads are defined by MRTD, with revenue increasing from $1.1 million in 2011 to $1.6 million in 2015. However, 30% of the toll revenue goes to paying for the operation of the toll stations and related staff, down from earlier collection costs amounting to 80% of revenue. The State Road Fund falls under the authority of RTDC and is under its Toll Collection, Road Use and Traffic Control Department, but is controlled by a Road Board consisting of representatives of road users and citizens. The composition and procedures of the Road Board are determined by MRTD. Unlike the previous Road Fund that was also used to finance road construction, the new State Road Fund only finances road maintenance and repairs in international and state roads, with priority given to maintenance, routine, and emergency repairs. Up to 10% of the State Road Fund revenue may be used to provide equipment to the state road maintenance companies.

Road Asset Management System

Development. Development partner support to RAMS development in Mongolia has been going on since 2009. Initially, data collection and the development of a simple MS Excel database was carried out by the then Department of Roads (DOR). Under the ADB technical assistance on Road Database Development Using Geographic Information System (2009–2011), survey equipment was provided, data were collected, a web-based database was developed, and annual and multi-annual plans were prepared. The subsequent Road Sector Capacity Development Project provided further support to data collection, further developed the road database, and prepared new annual and multi-annual plans. The support from ADB is currently being continued under the Regional Road Development and Maintenance Project with capacity building in the use of the RAMS for planning and prioritization.

Data collection equipment. Under the ADB technical assistance on Road Database Development Using Geographic Information System, a set of Road Measurement Data Acquisition System (ROMDAS) survey equipment was provided to DOR under the previous Ministry of Road, Transportation, Construction and Urban Development. The ROMDAS equipment included a bump integrator (IRI), a distance measuring instrument (chainage), video camera, GPS receiver, and laptop with ROMDAS data acquisition software. The ROMDAS equipment was again used under the ADB Road Sector Capacity Development Project to collect road network data. It was later transferred to the RTDC, but is no longer used for road network data collection.

Data collection. In 2003–2004, road surveys were carried out by the DOR for 3,809 km of paved state roads. Further data updates were provided in 2005–2008 for 688 km of newly constructed roads and 756 km of existing roads. Data was also collected for a total of 339 bridges and 2,172 culverts, including photographs. The technical assistance on Road Database Development Using Geographic Information System collected data for 6,411 km of state roads in 2010 using the ROMDAS survey equipment provided under the same

---

[31] The 1995 Law on Gasoline and Diesel Tax sets the fuel tax at MNT25,700 per ton for petrol and MNT2,140 per ton for diesel. The percentage allocated to the Road Fund is defined by the government on an annual basis.

[32] The revised Road Law does not mention the fuel tax, but earmarking of the fuel tax continues to exist in the Law on Petrol and Diesel Tax.

project. The collected data included GPS data (road track, intersections, bridges, culverts), pavement type, road signs, visual surface distress, roughness, and video. Data for other state roads and for local roads was obtained from GPS and other data collected by DOR. Traffic data were obtained from DOR for 71 counting stations (collected four times a year) for the period 2006–2009, and linked to GPS coordinates where the traffic counts were carried out.

In 2012 and 2013, data were updated under the ADB Road Sector Capacity Development Project using the same ROMDAS survey equipment. However, the data were not updated after the end of the project and the ROMDAS survey equipment is no longer used except for verifying construction quality, reportedly due to the cost of data collection. Some data continue to be collected and checked by the road maintenance companies. The 2017 revised Road Law defines that the state and private road maintenance companies are responsible for updating the road registration database and collecting traffic data (as well as preventing overloading).

Database. The revised Road Law defines that the road registration database should contain data on (i) the technical and operational parameters of the road; (ii) its capacity and status; (iii) asset value and depreciation; (iv) year of construction; (v) information on required repairs; (vi) information on damages and traffic accidents; and (vii) information on the entity that constructed, maintained, and repaired the road. This database is to be managed by the state central administrative body in charge of road matters, in this case RTDC.

An MS Excel database for the road network was developed in 2004, consisting of 25 data fields with data reported per 1 km section in the form of a strip map. The data includes gradient, curvature, surface type, left- and right-hand strip map, corridor width, pavement width, shoulder width, surface type, base course, soil embankment, bridges and culverts, snow risk, sandy or marshy areas, estimated speed, quality of pavement, roughness, skid resistance, pavement strength, traffic volume, left shoulder width, right shoulder width, sight distance, construction year, and original cost. The database contained data for approximately 5,200 km of roads collected in 2003–2004 and updated in subsequent years. The bridge and culvert survey data and related photographs were also entered into separate MS Excel files (one for each bridge or culvert).

Under the ADB technical assistance on Road Database Development Using Geographic Information System, a web-based database was developed using open-source PostgreSQL and PostGIS. Open-source Quantum GIS software was used for developing the GIS maps, and open-source Mapguide software and licensed Manifold software were used to develop the web interface. The PostgreSQL database included dynamic tables for Deighton Total Infrastructure Management System (dTIMS) data, accident data, traffic data, condition data, and weather data. These dynamic tables could be exported for use in MS Access or MS Excel, and subsequently re-imported into the database. Insufficient attention was given to institutionalization of the database, and at the end of the project DOR was unable to operate the database as the two trained staff members had left and the DOR servers had not yet been installed.

Under the ADB Road Sector Capacity Development Project, the existing database was updated and provided with a Mongolian language front-end to facilitate data entry. The new database was completed in 2014, and data from the 2012 and 2013 surveys were entered. However, the database was no longer used after the end of the project, and MRTD reverted to using the former MS Excel database with data checked and updated annually by the road maintenance companies. The RAMS database is currently being further developed under the Regional Road Development and Maintenance Project. This is making use of the commercial PAVER Pavement Maintenance Management System software. The software and related database are planned to be implemented by 2022.

Data analysis. Under the ADB technical assistance on Road Database Development Using Geographic Information System, a software license for dTIMS was provided to DOR. This was used to prepare a 5-year rolling plan for the period 2012–2018. The 5-year plan prioritized 12,624 km of regular repairs (overlays and seals) and 176 km of major repairs for the planning period, with a total cost of $151 million. Training in the use of the database and dTIMS software for planning was provided to staff of the Mongolian University of Science and Technology, given that DOR staff lacked the required technical knowledge and language skills.

Under the ADB Road Sector Capacity Development Project, a Mongolian front end was developed for dTIMS, facilitating its use by staff of DOR. Annual works programs were prepared for 2013 and 2014, including significant regular repairs to important corridors. However, the regular repairs were not approved by the Ministry of Finance, as has been the case for all requests for financing regular repairs submitted by MRTD. Since MRTD only receives funding for maintenance and routine repairs, they are especially interested in a RAMS that can help them plan these types of works. The dTIMS software, as most pavement management systems, focuses on the planning of regular and major repairs, and has therefore fallen into disuse by MRTD in favor of using the simpler MS Excel file that it was using before. Under the Regional Road Development and Maintenance Project, it was recommended to develop a type of decision matrix (treatment table) using dTIMS or HDM4, which would be updated every 5–10 years by specialized consultants, and which would be used to develop annual works programs based on a more limited set of data (e.g., only visual inspections).

A pavement management system developed by the Snowy Mountains Engineering Corporation (SMEC) was provided to the Ulaanbaatar Department of Roads (UBDOR) that is responsible for 280 km of main roads in the capital city. However, this rapidly fell into disuse, reportedly because of the high license fees for its use. UBDOR is currently using MS Excel for their data analysis. The prioritization is based on a "worst-first" approach, meaning that good and fair roads receive insufficient attention and quickly deteriorate.

Under the Regional Road Development and Maintenance Project, the commercial PAVER Pavement Maintenance Management System software developed by Colorado State University is currently being introduced. PAVER makes use of a pavement condition index (PCI) with a rating between 0 and 100 to assess the pavement condition and determine treatment options. This software is relatively inexpensive, but provides licenses for only a limited number of users (standard 2 users, expandable to 10 users). The PAVER software may be used with different types of databases, including MS Access, SQL, and local databases, and allows a single database to be accessed by different users.

Performance-Based Road Maintenance Contracts

In the case of the international and state roads, maintenance and repairs are contracted out to the 20 state road maintenance companies (AZZAs). The contracts are directly awarded for 1 year without competition. Many of these state road maintenance companies are very small, with annual turnovers in the order of $100,000–$200,000. These companies are inefficient due to high overhead costs and limited use of their equipment. There is a tendency to create new state road maintenance companies as newly improved roads are put under maintenance. However, the ADB Road Sector Capacity Development Project advised that this approach was inappropriate, and that consolidation of the existing companies was required to ensure their financial viability. Apart from the contracts with the 20 AZZAs, there are also 12 road maintenance and repair contracts with private companies (some of these involve previously privatized AZZAs). This used to involve 4-year contracts with annual renewals and work volumes determined on an annual basis. However, from 2021, these contracts are tendered out competitively on an annual basis for a period of 1 year. The different road maintenance and repair contracts with the AZZAs and private companies are generally area-based and include several different roads, ranging from 80–1,285 km in size, with contracts generally limited to a specific province or portion of a province. However, some contracts are more corridor-based and extend beyond the boundaries of a specific province.

Pilots. MRTD has indicated that it wishes to move to outsourced contracting for all road maintenance and routine repair contracts on the basis of PBCs. However, PBCs in Mongolia face a number of obstacles. While funding projections beyond 1 year are possible, it is not possible to commit funding to multi-annual contracts when government funding is used, even where funding from the State Road Fund is concerned. The use of multi-annual PBCs would, therefore, require a change to the Budget Law or a modification of the way the State Road Fund may be used. Under the existing legislation, contract amounts for government-funded contracts cannot exceed the cost estimate by more than 5%. Cost estimates are to be based on measured work quantities rather than compliance with performance standards, complicating adherence to this requirement in the case of PBCs.

Despite various obstacles for applying PBCs under government funding, these do not apply to development partner-funded pilots. Under the ADB Regional Road Development and Maintenance Project, two pilot PBCs were tendered in early 2020 for the road Ulaanbaatar–Arvaikheer (30.4 km and 27.2 km). The contracts had an estimated cost of $7.7 million each and included full rehabilitation (20 months) and subsequent maintenance and routine repairs (3 years). Initial rehabilitation was to be paid on an output basis, with subsequent routine and winter maintenance to be paid on a performance basis. The tenders were not successful, however, and a second tender was issued late 2020 including one single contract for the full 57.6 km and an extended rehabilitation period of 24 months. This was successfully awarded to a Mongolian company in April 2021 with a contract value of $14.0 million (equivalent to $49,000/km/year including the initial rehabilitation costs).

Performance standards. Under the ADB Road Sector Capacity Development Project, levels of service were defined. These distinguished four levels of service for international and state roads (urban or rural) and for capital city and local roads (main or other). The levels of service differed in terms of the response times set for correcting various types of defects. Under the ADB Regional Road Development and Maintenance Project, it was recommended to change the service levels to define maximum allowable defects instead of response times as this would be easier to manage and monitor. In the resulting set of performance standards applied in the pilot PBC, response times are only used in the case of events with a sudden onset that cannot be properly predicted or prevented (winter maintenance, emergency response, accident damage). For the other performance standards, the specifications clearly state that the contractor is required to address the different defects before the threshold is reached, and that deductions will be applied immediately where noncompliance is encountered during monthly inspections. The only exceptions are for the repair of signposts and road markings where response times continue to be used, complicating the inspection process and the calculation of payments.

The pilot PBC includes 15 separate performance standards, each with a set of underlying performance indicators and thresholds. The performance standards are defined in detail, explaining the activities to be carried out by the contractor and the related performance indicators and thresholds to be complied with. Not all defined activities are properly covered by the performance standards, however, complicating enforcement through the application of payment deductions in case such activities are not carried out.

Although the general specifications allow inspections to be carried out for a sample of the contracted road length, for the pilot, the requirement is that the entire road length is inspected each month. During the inspections, compliance is assessed for each 1 km section or for the surveyed road section as a whole, depending on the performance standard. A network performance score (NPS) is subsequently calculated by multiplying the percentage of the surveyed road length that is noncompliant, with a weight for the specific performance standard and by a multiplication factor, and subtracting the resulting percentage from 100% (representing full compliance). The resulting NPS ranges from 0%–100% and is multiplied by the monthly payment amount to determine the actual payment for a specific month. This provides a clear and transparent system that is easy to understand by all parties.

The weights for the performance standards in the pilot PBC differ for the two lots, as one lot does not include any bridges. Weights are higher for pavement, shoulder, vegetation control, and winter maintenance. Despite its importance, the weight for drainage is relatively low, equivalent to that for marker posts. Together, the weights of the 15 performance standards add up to 100%, implying that only full noncompliance with all performance standards for the entire road length would result in the deduction of the full monthly payment. The multiplication factor improves this situation slightly by increasing the deduction by a factor 1.5 for most performance standards, and a factor 5 in the case of winter maintenance (the use of the factor corresponds to an increase in the weights being applied to each performance standard). However, in summer months, this still means that the contractor would receive no payment only in case of noncompliance for all performance standards for at least 80% of the road length. To avoid such a degree of noncompliance, the contract stipulates that the contract may be terminated if the NPS repeatedly falls below 50%. However, an NPS of 50% still requires noncompliance with all performance standards for at least 40% of the road length. In practice, with a newly rehabilitated road, this implies the contractor can do nothing for the full 3-year maintenance period and will still be eligible for payment. With the contract only recently signed and the performance-based maintenance only starting after the rehabilitation works have been fully completed, it will still take some time before the suitability of the performance standards and related deduction calculations can be properly assessed.

Way Forward

Road asset management system. Mongolia has a long history of RAMS development. This started with a simple MS Excel system used mainly for routine maintenance planning, which continues to be used today. More comprehensive systems that were developed with ADB support have been used to prepare regular repair programs, but these programs have not been funded by the government. Without such funding for regular repairs, the added value of the comprehensive systems does not weigh up against the additional needs for data collection and system operation. As a result, these comprehensive systems have repeatedly fallen into disuse. There is currently a need for a more basic RAMS that is light on data collection and system operation needs (similar to the MS Excel system), but that has the ability to also identify and prioritize regular and major repair needs. This will still need to be web-based, but should be able to function on a minimum set of data that is easy and inexpensive to collect. The planning module needs to be transparent and should be based on a decision matrix to be prepared by consultants and agreed with government. At the same time, there is an urgent need to convince the government of the need to allocate funding to regular repair in order to reduce the deterioration of the road network and to avoid the need for costly major repair. Unless the funding issue can be solved, the benefits of an improved RAMS will not become evident.

Performance-based road maintenance contracts. Mongolia has only recently awarded its first PBC, and the implementation of the performance-based maintenance and related performance payments will only start in 2 years after rehabilitation has been completed. It will be important to carry out a proper evaluation of this pilot as the basis for further replication. The performance standards being applied are very comprehensive and incorporate several lessons learned from other countries, especially related to the use of response times. However, further improvement of these performance standards is required to ensure that they properly cover the different activities required under the contract. Most importantly, further attention needs to be given to the calculation of the payment deductions, ensuring that this provides appropriate incentives to the contractor to comply with the performance standards. Attention also needs to be given to developing a suitable legal framework for application of PBCs under government funding, allowing the approach to be replicated by government in future years.

Pakistan

Road network. Pakistan has a road network of 263,775 km, of which 13,570 km are defined as national highways, including 78 km of expressways, 2,066 km of motorways, and 11,426 km of highways. Although the national highways make up less than 5% of the road network, they carry over 80% of the traffic. The 2005 National Highway Authority (NHA) Code and Standard Operating Procedures for the Road Maintenance Account distinguish between routine maintenance (pothole patching, crack sealing, drainage cleaning, vegetation control, etc.); periodic maintenance (chip seals, asphalt overlays, structure repairs, etc.); emergency maintenance (ensuring safety and traffic flow after landslides, flooding, heavy snowfall, and accidents); rehabilitation (thick overlays, replacement of structural layers, major repairs, or replacement of structures); and so-called geometric improvements (involving reconstruction with changes to width, curvature, alignment, and gradient of roads and structures).

Institutional framework. The national highways are managed by the NHA, which was created in 1991 through an act of Parliament. NHA is responsible for planning, development, operation, repair, and maintenance of national highways and strategic roads entrusted to NHA by the federal government or by a provincial government. It is managed by a National Highways Council and an Executive Board. In 2000, the National Highway Council created a Road Asset Management Division (RAMD). This was initially under the Operations Wing of NHA that also includes the 10 regional maintenance divisions, but was later placed under the Planning Wing. The responsibilities of RAMD have expanded over the years, and the RAMS is currently operated by the RAMS Directorate under RAMD. RAMD also manages the road maintenance account (RMA) that is financed from road user charges and provides funding for road maintenance and rehabilitation. Actual collection of road user charge revenues (tolling, weigh stations, and use of the right-of-way) is carried out by the Finance Wing of the NHA. All road works are tendered out competitively to the private sector, as are the contracts for road tolling and data collection. A RAMS Working Group was recently set up with participation from NHA as well as from the different provinces. The RAMS Working Group supports the exchange of experiences, allowing provinces to further strengthen their development of RAMS.

Financing. The World Bank supported the creation of a Road Fund with off-budget earmarked funding sources since the late 1990s. Initially, the aim was to create a second-generation Road Fund to finance both national and provincial roads, financed from a fuel levy and managed by a Roads Board. However, this was rejected as it was expected to lead to issues regarding the distribution of revenue between the different provinces. Instead, in 2003 an RMA was created under NHA focused only on national highways, established through a ministerial notification and funded primarily from tolls on the national highway network. NHA was already entitled to collect tolls under the NHA Code, and simply earmarked these to the RMA together with other road user charge revenues collected by it. Despite having a more fragile status than a Road Fund, it has been very successful in fulfilling its objectives. Similar road funds have been created at provincial level for funding provincial highways.

Revenues for the RMA include tolls on roads and bridges, traffic fines, licenses for use of the right-of-way, axle-load charges, heavy-vehicle fees, international transit fees, and border fees. These road user charges are complemented by annual maintenance grants from the federal government, funds provided by development partners, and loans obtained by NHA. The RMA revenue amounted to approximately $70 million upon creation in 2003, up from the federal maintenance allocation of $14 million the year before. RMA revenue quickly increased in subsequent years, reaching approximately $200 million in 2011–2012. For 2020–2021,

RMA revenue is expected to reach approximately $250 million (with the total NHA budget expected to reach approximately $780 million). Despite this continued growth in maintenance funding, maintenance needs have been increasing even more rapidly, resulting in a maintenance backlog and an estimated maintenance need of $375 million in 2019–2020. In part, this is the result of many provincial highways being reclassified as national highways and falling under the management of NHA. The funds in the RMA are to be used in order of priority for (i) routine and periodic maintenance, (ii) rehabilitation, (iii) geometric improvements (max 6%) and road safety improvements (max 5%), (iv) new toll plazas and weigh stations (max 2.5%), and (v) corridor management (max 1.5%). The use of the RMA is managed by the RAMD.

Road user charges. The NHA collects several road user charges that are earmarked to the RMA. Tolls were first introduced in 1999, and in 2003 toll revenue was earmarked to the RMA. Toll revenue continues to be the most important source of financing for the RMA, increasing significantly in the initial years from approximately $35 million in 2003 to $116 million in 2011. Of the $250 million in RMA revenue expected for 2020–2021, toll revenue is expected to make up $145 million. The increase in toll revenue is in part the result of additional toll stations, but is also due to a significant increase in toll rates in 2005. Although toll rates continue to be very low compared to other countries, low usage of toll roads means that further rate increases are not realistic. Toll collection is tendered competitively, with bids either defining the fixed monthly payments to NHA against the right to collect tolls at predefined rates, or defining the percentage of toll revenue to be retained by the bidder. The percentage of RMA revenue made up by tolls has reduced from over 90% in 2005 to less than 60% for 2021, as revenues from other road user charges have also increased. Other important road user charge revenues include traffic fines and licenses for the use of the right-of-way. For the latter, the revenue has been limited as licenses issued before the creation of NHA are generally long term (99 years) and provide little revenue. NHA is currently focusing efforts on identifying all usage of the right-of-way so that this may lead to increased revenue.

Road Asset Management System

Development. RAMS development in Pakistan has been supported by development partners, especially the World Bank and ADB, for the past 20–30 years. This initially focused on the data collection and RAMS development for the national highways managed by NHA, leading to the successful institutionalization of the RAMS in NHA annual maintenance and rehabilitation planning. As the NHA RAMS became sustainable, the attention of the development partners shifted to developing the RAMS for the highway and road authorities in the different provinces. This is largely based on the NHA RAMS, and makes use of similar setups for data collection, management, and analysis. However, maintenance funding is more of a challenge at provincial level as traffic volumes are much lower on provincial highways and tolling is therefore less appropriate.

Data collection equipment. NHA have their own data collection equipment. This started with the procurement of Road Measurement Data Acquisition System (ROMDAS) survey equipment including a GPS receiver, distance measuring instrument (DMI), and a laser profilometer and bump integrator to measure road roughness (IRI). A trailer-based Dynatest heavy FWD was later procured, followed by the recent procurement of an LCMS for automatically assessing the degree and severity of cracking. The survey equipment was initially used in-house by NHA, but is currently provided to consultants as part of the outsourced road condition survey contracts.

Data collection. Road condition data are collected between 1 July and 15 September every year. The minimum data to be collected is defined in the 2005 Standard Operating Procedures for the RMA. This includes inventory data for roads (length, width, surface type, number of lanes, shoulders, guardrails, etc.); bridges (location, length, type, load rating, etc.); and traffic signs (location, type, size, support, etc.).

Pavement condition data includes the percentage of the road length affected by cracking (including crack width), rutting (including rut depth), potholes (including the number of potholes), raveling, edge step, and edge break, as well as roughness (IRI). Structural capacity is also measured using FWD and surface friction is measured using a skid resistance measuring device. Traffic data are collected by means of 24-hour, 7-day traffic counts in fixed locations in the network. The traffic counts are increasingly being complemented by data collected from the tolling stations. Additional traffic data are collected regarding the vehicle fleet (number and type of vehicles and basic characteristics) and the truck loads (from weigh station data). Survey data are complemented by secondary data on road user costs and treatment costs per kilometer (for use in HDM4).

Data used to be collected in-house by NHA, but currently, collection of condition data is outsourced, covering 3 years of condition data collection for the entire road network and the subsequent data analysis and preparation of annual maintenance plans. The current contract is for financial years 2018 to 2020, and a tender for a new 3-year contract was published in 2020. The data collection includes information on the extent and severity of surface distress (rutting, cracking, potholes, raveling, edge drops, edge breaks); roughness; pavement bearing capacity and remaining life; and drainage condition. The consultant is provided with the NHA survey equipment including a distance measuring instrument, laser profilometer, bump integrator, LCMS, and FWD. The consultants are responsible for calibration and maintenance of the equipment. The consultants are also required to provide geotagged photographs for each kilometer section of road and for each major bridge.

In 2016, NHA decided to collect expanded inventory data for its network and initiated the procurement of consultants to collect GPS inventory data for the entire national highway network. Due to problems with the procurement, the contract award was only finalized mid-2019, with the contract going to a joint venture of two Pakistani companies for just under $1 million. The contract includes the collection of geotagged inventory data for the road alignment divided into 1 km segments (including data on length, number of lanes, carriageway width, surface type, medians, shoulders, etc.), complemented by geotagged inventory data on all road signs (type, dimensions, condition); road structures and furniture (location, type, dimensions and condition of retaining walls, guardrails, fences, parapets, headwalls, etc.); GPS demarcation of the right-of-way; geotagging of any utilities within the right-of-way; and geotagging of all other public or private assets within the right-of-way (toll plazas, weigh stations, rest areas, fuel stations, intersections, commercial or private properties, etc.). For each road asset, geotagged photographs are to be collected as well as high-definition video of the road itself. The surveys are to be complemented with the integration of existing data already with NHA. An important objective of the data collection is to identify all usage of the right-of-way and increase the revenue collected from licensing its use.

At provincial level, a similar approach is being applied. In Sindh Province, for instance, consultants are currently involved in the survey of 57,000 km of roads for the Works and Services Department of the Provincial Government, funded with ADB support.

Database. In the early 1990s, a maintenance management system was introduced that, over time, evolved into the RAMS. The 2005 Standard Operating Procedures for the RMA call for a centralized database with a local area network (LAN) connection, to be later upgraded to a wide area network (WAN) or possibly an internet connection. Support to the development of an upgraded RAMS was provided under the World Bank Highway Rehabilitation Project, consisting of a database and a pavement management system.

In 2015, NHA expanded the RAMS to include a bridge management system. In 2017, a road safety module (PakRAP) was added, based on the integrated road assessment program (iRAP). In 2019, NHA decided to expand the functionality of the RAMS and its different modules by adding GIS functionality, and it contracted the collection of geotagged inventory data together with the development of a centralized geo-database using ESRI Enterprise ArcSDE software, complemented by a GIS web application to allow access through the internet. All data are to be stored in the geo-database, with the web application providing an easy-to-use interface to generate standardized reports, carry out basic analysis, and prepare maps. The consultants are also required to provide training to NHA users, system administrators, and management staff.

Data analysis. The 2005 Standard Operating Procedures for the RMA define that HDM4 will be used as the decision support system for the analysis of the road network data. The data analysis is carried out in September and October each year, in preparation for the annual budgeting. This starts with an HDM4 strategy analysis to determine the funding needs, which is submitted to the NHA chairperson and executive board together with recommendations on changes to the road user charges and related resource mobilization. HDM4 is also used to carry out a program analysis to determine the priority road sections. A long list of roads is first prepared, taking into account the priorities defined by the NHA regional offices. Using HDM4, an unconstrained work program is prepared to determine total funding needs. Based on the expected funding levels, a constrained work program is subsequently prepared, and a short list of roads is identified for inclusion in the Annual Maintenance Plan (AMP). The AMP includes a Routine Maintenance Plan covering the entire NHA road network, as well as a prioritized Periodic Maintenance Plan and Rehabilitation Plan based on the results of the HDM4 analysis. Although priorities are defined based on economic analysis (net present value per unit of cost), up to one-third of the value of the plan may be amended based on regional equity considerations (as long as these have a positive net present value/cost ratio). Roads that cannot be included in the AMP, are included in the rolling 5-year plan. The AMP is subsequently discussed with the NHA regional offices between October and November, and detailed cost estimates are prepared between November and the end of December. The updated AMP is subsequently reviewed by the RMA Technical Scrutiny Party and the RMA Steering Committee, before approval of the final AMP by the NHA chairperson or executive board by 31 January. Bidding documents are subsequently prepared in February and March and contracts are procured and awarded by 31 May, allowing new maintenance contracts to start in the new financial year (1 July–30 June).

The data analysis used to be carried out in-house by NHA, but in recent years the HDM4 analysis and more recently the preparation of the AMP itself is included in the 3-year contracts for road condition surveys. The contracts clearly define that the consultant is to use HDM4 strategy and program analyses for the preparation of the AMPs. The consultants are also to provide training to NHA staff in the use of HDM4.

Coverage of long-term maintenance needs by the RMA increased from 25% in 2003 to over 80% in 2011, but has since decreased again as maintenance costs and the size of the network managed by NHA have increased.[33] The use of the RAMS in the preparation of the AMPs has resulted in a shift from rehabilitation of poor condition roads, toward routine and periodic maintenance of roads in good and fair condition. In 2003, only 31% of AMP funding was going to routine and periodic maintenance, but by 2005 this increased to 70%. Periodic maintenance has especially increased significantly, making up most of the AMP budget. Initially, there was a lack of contractors with the required skills and equipment to carry out the periodic maintenance contracts, but contractors have been quick to adapt to the new demand.

At provincial level, HDM4 is also being used. In Sindh Province, the consultants currently hired with ADB support are using HDM4 to determine treatment needs and generate a priority list of maintenance works to be reflected in a Road Maintenance Management Master Plan that is to be prepared for the Works and Services Department of the province.

---

[33] A number of provincial highways were reclassified as national highways, many of which require rehabilitation.

Performance-Based Road Maintenance Contracts

Pilots. Development partners have been promoting performance-based contracting for road maintenance in Pakistan for the past 2 decades. However, no PBCs have been piloted yet. In part, the lack of PBC pilots is due to the NHA Code that stipulates that all routine maintenance contracts will be output-based measurement contracts based on unit rates from the Composite Schedule of Rates (periodic maintenance can be contracted on a lump-sum basis). Applying performance-based maintenance is, therefore, only possible under concession contracts such as the four motorway concession contracts. These are build–operate–transfer concession contracts for 10–25 years, covering a total length of 584 km and involving a total cost of approximately $1 billion, with investments recovered through tolling.

Under the ADB-funded Khyber Pakhtunkhwa Provincial Roads Improvement Project, two 5-year PBCs are currently being prepared, covering 104 km of provincial roads that are in good to fair condition. This will focus on maintenance with only limited initial repairs. The pilots will be fully financed with funding from the Kyhber Pakhtunkhwa Highway Authority. The project budget is $10 million, implying an average cost in the order of $20,000/km/year.

Under the recent ADB Enabling Economic Corridors through Sustainable Transport Sector Development technical assistance, an approach to management agency contracts (MACs) was proposed. This involves a PBC covering a portion of the road network, resulting in larger contracts likely to attract more experienced contractors. The resulting guidelines recommend MACs for contracts between the NHA or provincial highway authorities and contractors, as well as for use between the federal/provincial government and the highway authorities. The guidelines include a sample contract document for use with a contractor. To avoid the problems with performance-based contracting of routine maintenance, the guidelines propose procurement of the MACs as concession contracts under Pakistan's Public–Private Partnership (PPP) Policy (maintain–operate–transfer, as opposed to build–operate–transfer). However, the PPP Policy is primarily aimed at motorways with ownership transferred to the concessionaire and financing provided through tolling and additional payments from government, as required. Although such operation and maintenance concession contracts are possible, they are more complicated than PBCs under traditional procurement and with agreed payments by government. Discussion of the PBCs and MACs in the RAMS Working Group created under NHA has led to greater interest in PBCs, with additional PBC pilots being discussed for a new project being prepared for Punjab Province. Sindh Province is also preparing PPP contracts for two roads for operation and maintenance over a period of 10 years. This includes the development of a set of performance standards and related bidding documents to be used in the PPP contracts.

Performance standards. As part of the MACs prepared under the ADB Enabling Economic Corridors through Sustainable Transport Sector Development, a set of service levels and related performance standards were also prepared. These distinguish 55 performance standards related to pavement, vegetation, cleanliness, roadside furniture, electrical systems, drainage, roadside earthworks, structures, traffic accidents, buildings, winter maintenance, and management reporting. The deduction system relies heavily on response times, depending on regular inspections to record any defects and follow-up inspections to assess its correction within the specified response time. Deductions of 0.1% of the monthly payment amount are applied for each day the response time is exceeded (defects affecting passability of the road have deductions applied every hour). This does not take account of the fact that many of these defects can be predicted and should be prevented from exceeding the allowable threshold rather than correcting them once the threshold has been exceeded. More importantly, the use of response times as the basis for deductions, means that inspections have to be frequent, daily or even hourly, to assess compliance by the contractor, resulting in a high supervision burden. The definition of deductions as a percentage of the total monthly payment does not take account of the length of road under contract, with contractors with larger networks and higher monthly payment amounts receiving a higher deduction for the same defect. This approach also does not take account of the number of noncompliance under a specific standard, with one pothole resulting in the same deduction as 20 or 200 potholes.

Way Forward

Road asset management system. Over the past few decades, the NHA has developed a sustainable RAMS which is fully institutionalized into annual planning and budgeting procedures. From a very basic database, it has evolved over the years into a RAMS with GIS mapping and expanded functionality. Data collection, management, and analysis used to be carried out in-house, but are now contracted out to the private sector as part of the gradual reform of NHA. The creation of the Road Maintenance Account and earmarking of road user charge revenues has allowed most maintenance needs to be financed. The RAMS in Pakistan has reached maturity, although further improvement and expansion of functionality will continue to be required in future years. At the provincial level, RAMS development is ongoing and will require continued support from development partners. A major challenge is the financing of provincial highway maintenance, with toll revenue being a less suitable source of funding due to the lower traffic levels on provincial highways. The introduction of a fuel levy should be revisited, with clear guidelines on the sharing of revenue between the different highway networks and related highway authorities.

Performance-based road maintenance contracts. Pakistan has not yet implemented any PBCs, although two pilots are currently under preparation and further pilots are being discussed. The NHA Code forms an important obstacle to the use of PBCs in national highways, requiring output-based contracting for routine maintenance. Although performance-based maintenance is possible under concession contracts regulated by the PPP Policy, this is a more advanced approach with higher requirements regarding financing and management that is not suitable for all highways managed by the NHA. An amendment of the NHA Code is recommended to allow for the use of PBCs. Suitable performance standards and related deduction formulas will need to be developed in support of future PBCs. These should be usable independently of the contract size and should not depend on overly frequent inspections. For the initial pilots, the applied performance standards and deduction calculations should be simple and transparent, facilitating understanding and avoiding disputes.

Tajikistan

Road network. Tajikistan has a road network of approximately 26,600 km, of which 14,339 km are public roads, including 3,348 km of international roads, 2,127 km of republican roads, and 8,864 km of local roads. In terms of road activities, distinction is made between maintenance (excluding pavement and structure repairs), current repair (pavement and structures), midterm repair (periodic maintenance), capital repair (rehabilitation), and (re)construction.

Institutional framework. The Ministry of Transport (MOT) is responsible for the management of the public roads. It is supported by six state institutes for road administration in the different regions. Routine and winter maintenance are carried out by the 64 state institutes for road maintenance (GUSADs) under MOT, while larger works are carried out by the State Unitary Enterprise, Directorate of Construction Enterprises under MOT or contracted out to contractors. Design, quality control, and diagnostics are carried out by the State Unitary Enterprise, Transport Design Institute, and the State Service for Supervision and Regulation under MOT. A Road and Transport Sector Digitization Unit was established in 2021 under MOT, which will be responsible for the operation of the RAMS that is currently under development.

Financing. Financing for the road sector consists of annual allocations from the Republican Budget and development partner loans and grants. In 2019, the road sector funding amounted to $186.6 million, equivalent to an average of $13,000 per kilometer of road. Road maintenance funding has been increasing in Tajikistan somoni terms, but has remained more or less constant in US dollar terms, making up only $7.3 million in 2019, equivalent to $500 per kilometer of road and forming only 0.1% of GDP. The MOT indicates that actual maintenance needs are approximately five times what they are currently receiving. This was corroborated by an HDM4 analysis carried out in 2017 that estimated a budget requirement of $32 million per year to stabilize the road network conditions. This is equivalent to 0.4% of GDP, in line with the minimum international benchmark for road maintenance allocations. Under the Obigarm–Nurobod Road Project, the government signed a covenant with the EBRD to set up a dedicated Road Maintenance Fund by the end of 2023 that will be solely used for road maintenance, and to increase road maintenance funding to $12 million per year by the end of 2025.

Road user charges. A corporate road user tax is collected from businesses based on turnover ($38 million in 2019), but is no longer earmarked for the road sector and is projected to be abolished in 2021. A fuel excise tax ($15 million in 2015) and a vehicle tax ($20 million in 2018) are collected, but are not allocated to the road sector. Tolls are collected on the concession road from Dushanbe through Khujand to the Uzbekistan border (approximately $10 million in 2019) and are applied by the concessionaire for the maintenance of the road (maintenance of the three large tunnels is not included). The government is currently planning to review actual maintenance needs and identify possible funding sources to finance the planned Road Maintenance Fund. This will include a review of options for introducing tolling projects by the end of 2024.

Road Asset Management System

Development. Data were collected and a database was prepared with ADB support in 2008, but this is no longer in use. A government decree was recently issued for a RAMS Action Program for the period up to the end of 2024. In line with the RAMS Action Program, MOT has established a Road and Transport Sector

Digitization Unit that will coordinate the data collection and the operation of the RAMS. The members of the Road and Transport Sector Digitization Unit will receive training and support under various ongoing development partner projects.

Data collection equipment. GUSADs carry out spring and autumn inspections, but this mainly involves visual surveys and data are not stored centrally. In the past, JICA has supported basic data collection in four regions[34] using the Dynamic Response Intelligent Monitoring System (DRIMS). This is a class III accelerometer attached above the rear wheel of a vehicle that collects roughness data and combines this with data from a basic handheld GPS device. Data was recorded in a laptop PC, which calculates IRI values for 100-meter sections. The accelerometer is calibrated by driving the vehicle over two standard humps at a predetermined speed. Video data are also collected using a dashcam, which was subsequently used to identify cracking and pothole ratings at 100-meter intervals.

The World Bank, under its Second Phase of the Central Asia Road Links Program, provided 40 fixed traffic counters, which are currently being installed at key locations. The counters use video to automatically classify vehicle types and are linked to a server in MOT. A Trassa road survey vehicle was also provided that collects roughness and chainage data, but does not collect GPS data. The data can only be exported in PDF format in order to avoid possible manipulation of survey data in case of quality control of completed works. However, this means that the collected data cannot easily be exported to the RAMS database.

ADB is planning to support the procurement of a network survey vehicle in 2021 under its ongoing CAREC Corridors 2, 5, and 6 (Dushanbe–Kurgonteppa) Road Project—Additional Financing. The network survey vehicle will enable collection of chainage, GPS, roughness, and video data, allowing for post-processing of video data for inventory and visual condition assessments, and enabling all data to be linked through the GPS data for visualization using GIS. The procurement of the network survey vehicle will be complemented with handheld devices (e.g., tablets) for recording localized inventory and condition data. Discussions are ongoing regarding the purchase of a light detection and ranging (LiDAR) system for the collection of detailed road inventory data in support of the updating of road passports. Although the data collection is relatively quick with this equipment, significant post-processing is required to transform the collected data into a usable format for the RAMS. Most of the collected data will also not be used in the RAMS data analysis and planning at network level.

Data collection. The JICA-funded technical assistance used the DRIMS equipment with the GUSADs in the regions and collected roughness data for approximately 1,300 km of roads in 2013, 2014, 2015, and 2016 (Gissar and Kurgontebba regions). Further network data collection has not yet taken place. The RAMS Action Program defines that data collection will be carried out with consultant support, focusing initially on international roads, and gradually expanding to also include republican and local roads. ADB is planning to support the data collection under its recent Road Network Sustainability Project in the period 2022–2023. Data collection will be carried out by consultants with significant involvement of MOT staff. According to the RAMS Action Program, after this initial support from project consultants to collect full inventory and condition data for the network and train MOT staff, annual data collection will be carried out by the MOT for 100% of international roads, 50% of republican roads, and 35% of local roads each year (average 7,500 km per year).

Database. A database was developed in 2008 under the ADB technical assistance Strengthening Implementation of the Road Maintenance Financing System. This Highway Information System (HIS) was an MS Access database with a Russian language interface. However, the database has not been updated since 2008 and is no longer in use.

---

[34] Mainly Gissar and Kurgontebba, and to a more limited extent in Sugd and Kulyab.

The World Bank, under its Second Phase of the Central Asia Road Links Program, supported the procurement of computer hardware and software for MOT, as well as the development of a tailor-made RAMS database by a local software developer. This includes an electronic road mapping module and a road database module. The database itself is structured around a set of 16 data collection forms that are based on old Soviet Government Standards (GOST),[35] and which do not necessarily meet the needs of a modern RAMS. These include data requirements for road code and name, general road data, road geometric parameters, visual road condition assessments, pavement roughness, road accidents, traffic volume, bus stops, service and rest areas, culverts, pavement structure, bridge and overpass inventory, bridge and overpass condition, repair works, strip maps, and tunnels. As such, the data requirements are quite extensive. Road data are stored by road link, rather than by segment. The database includes a number of standard reports, but these are very rigid and hard-programmed into the database, limiting its use for road asset management. The development has lacked support from experienced road management specialists, and the resulting database lacks the necessary functionality for road asset management.

Support to the development of the RAMS database will be continued under the new World Bank-funded Fourth Phase of the Central Asia Road Links Program, and will include development of a GIS module for mapping the collected data. This will build on the existing database, expanding the additional functionality for road asset management. The project is coordinating with ADB's Road Network Sustainability Project that will support data collection and validation, allowing the database development to be adjusted to the data storage requirement and enabling the testing of the database. According to the RAMS Action Program, the initial database development and testing will be followed by a period of further testing and finalization of the database by 2023. From 2023 onward, the Road and Transport Sector Digitization Unit under the MOT will be fully responsible for operating the database.

Data analysis. Under the Second Phase of the Central Asia Road Links Program, support has also been provided to the development of a set of algorithms for predicting road deterioration and prioritizing budget allocations to different roads and treatment types. These are reportedly based on HDM4 algorithms that have been simplified to reduce the data needs. However, the resulting algorithms have not been tested and it is unclear how well they will perform in the context of Tajikistan. With a lack of historical pavement performance data, it will also be impossible to test the algorithms and compare them to actual deterioration. The use of such tailor-made algorithms therefore introduces significant risks. This is even more the case since the algorithms appear to be programmed into the planning module, with limited scope for adjusting and calibrating them. It also means that there is a lack of support services and experience from users in other countries that can be used as a basis for testing and calibrating the system in Tajikistan. For a country just starting to introduce road asset management, this is not a recommended approach. The resulting algorithms are therefore not considered very suitable for incorporation into a planning module for the RAMS.

ADB's Road Network Sustainability Project is planning to carry out an HDM4 analysis of the collected data and to compare the results with the algorithms as part of the calibration process. Based on the results of this comparison, it may be necessary to replace the developed algorithms with an alternative planning module. This may be based on a decision matrix, which could be prepared using the results from the HDM4 analysis. The HDM4 analysis will also be used to prepare annual plans and 3-year rolling plans. According to the RAMS Action Program, the planning will be carried out directly by MOT from 2023 onward.

---

[35] государственный стандарт (ГОСТ).

Performance-Based Road Maintenance Contracts

Pilots. The government has carried out four PBC pilots. These pilot projects were financed by the government as part of project counterpart funding. The first two pilots were carried out under the ADB-funded CAREC Regional Road Corridor Improvement Project. This project was originally intended to cover five PBC pilots in Tajikistan as well as five pilots in the Kyrgyz Republic, but in the end only two pilots were carried out and only in Tajikistan. The PBC pilots started in 2013 and included two sections of the international road from Dushanbe to Karamyk (73 km from Vahdat to Obigarm and 76 km from Nurobod to Nimich). These road sections had been rehabilitated 6–8 years previously and required some initial repairs. The pilots involved output- and performance-based road contracts (OPRCs) with initial repair works paid on an output basis (volume of work completed), followed by 3 years of routine maintenance paid on a performance basis. Winter maintenance was paid on an output basis against work orders. The two contracts were awarded to domestic contractors against an average cost of $7,600/km/year including initial repairs, as well as routine, winter, and emergency maintenance ($5,800/km/year excluding initial repairs).

A second set of PBC pilots was carried out under the ADB-funded CAREC Corridors 3 and 5 Enhancement Project. The contracts started in 2018 and included a section of the same international road from Dushanbe to Karamyk (89 km from Sayron to Karamyk) and a section of republican road (87 km from Vose to Khovaling). These road sections had been recently rehabilitated, and the contracts therefore did not include initial repairs and were limited to 3 years of performance-based routine maintenance and output-based winter maintenance. Both contracts were awarded to domestic contractors. The average costs under these contracts decreased to $1,500/km/year, reflecting an increased understanding of the PBC concept.

Performance standards. For the performance-based routine maintenance, use is made of a set of performance standards indicating the maximum threshold for different types of defects. The performance standards are linked to a response time to correct the defect and a to a deduction to the monthly payment in case of noncompliance. A review of the PBC pilots identified a few issues with the current performance standards. One of the areas for improvement is to remove the response times for contractors to correct identified defects, as these are causing contractors to only carry out the works identified in the inspections as noncompliant, rather than proactively correcting all defects. The thresholds for some performance standards are very high (e.g., pothole sizes of up to 0.5 m², equivalent to a diameter of 80 cm) and need to be corrected to avoid unacceptable damages. The deductions to monthly payments in case of noncompliance were found to be very low, and in no relation to the cost of carrying out the repairs or to the potential cost to road users (e.g., deductions amounting to $5).

Way Forward

Road asset management system. The coming years will see the implementation of the RAMS Action Program, including the collection of network data for a large portion of the trunk road network and the development of a RAMS database and related data analysis tools with support from different development partners. The main challenge will be to ensure coordination between the different projects supporting RAMS development, and to achieve the integration of the RAMS into existing government procedures and budgeting systems. The continuous involvement of the Road and Transport Sector Digitization Unit under the MOT will be crucial in this respect, ensuring a transfer of skills and experience. The annual allocation of government funding for data collection and operation of the RAMS will also be a requirement to ensure that the RAMS will continue to function when the technical assistance ends.

Performance-based road maintenance contracts. Further review of the PBC pilots is required to develop suitable PBCs for application in Tajikistan. This will need to take into account the lessons learned from the initial pilots regarding the performance standards and response times, but should also include an extension of the contracting period to at least 5 years to ensure that this includes a few years with increased maintenance needs (this may then be followed by a new PBC including midterm repair). This next piloting period will also need to look at the integration of PBCs into government systems. This will face difficulties in terms of replacing the current approach of direct budget allocations to GUSADs and requiring multi-annual budget commitments. The future role of GUSADs will also need to be determined. This may involve a type of performance-based service-level agreement (SLA) applied to the direct contract agreements with GUSADs, but may go further and include reform of GUSADs and opening up of maintenance contracts in a portion of the road network to competition with participation of the private sector. Existing legislation and standards may complicate the use of PBCs or the involvement of contractors for government-funded repair and maintenance work, as they have in other countries.

Turkmenistan

Road network. According to the Ministry of Finance and Economy, Turkmenistan currently has 13,773 km of main roads, including 6,514 km of state roads (2,284 km of which are international roads) and 7,259 km of local roads. In terms of road activities, distinction is made between maintenance (excluding pavement and structure repairs), current repair (pavement and structures), midterm repair (periodic maintenance), capital repair (rehabilitation), and (re)construction.

Institutional framework. Until recently, the main roads were managed by the State Concern Türkmenawtoýollary that reported directly to the Cabinet of Ministers and carried out all construction, repairs, and maintenance in-house through its subordinate road organizations and outsourced specialized companies. In January 2019, Türkmenawtoýollary was merged with the Ministry of Construction and Architecture that is responsible for determining the road sector policies through its Highway Management Department. The responsibility for the construction, repair and maintenance of the state and local road network has been transferred to the administrations of the five regions and the capital city Ashgabat that act as the client for these roads. The former subordinate road organizations of Türkmenawtoýollary have also been transferred to the five regions and Ashgabat city.

Financing. Financing for the main roads comes from annual budget allocations from the state budget. There is no earmarking of revenues for the road sector.

Road user charges. The government collects various road user charges. A fuel excise tax of 40% of the cost is levied on both petrol and diesel. There is an excise tax on passenger cars equivalent to $0.3 per cubic centimeter of engine displacement. A vehicle ownership fee is levied annually and proof of payment needs to be presented at the technical inspection every 2 years. The vehicle ownership fee is based on a fixed rate, with passenger cars paying double the rate, buses three times the rate, and trucks six times the rate. None of the road user charges are earmarked for the road sector and instead go to the general state budget.

The government is further planning to construct a number of new toll roads. After initial difficulties with the international contractor for the planned 564 km toll road between Ashgabat and the Caspian seaport of Turkmenbashy, a consortium of domestic private companies was set up as closed joint stock company Turkmen Awtoban, which is responsible for designing and constructing the 600 km toll road from Ashgabat to the second-largest city Turkmenabat.

Road Asset Management System

Development. The road sector receives very limited support from development partners and is largely financed by the government and domestic banks. The government has not yet initiated any RAMS development in the country, and data collection is limited to basic data on the road names, codes, and lengths, as well as annual visual assessments of road conditions.

Data collection equipment. No specific data collection equipment is currently in use, and inventory and condition data are collected by means of visual assessments.

Data collection. Basic data on the state and local roads is collected on an annual basis by the administrations of the five regions and Ashgabat for the state and local roads under their jurisdiction. Collected data includes the road name, road code, length, type of permitted use, technical category, location, book value, and depreciation. This is complemented by annual visual assessments of road conditions and road maintenance needs by the state-owned road enterprises.

Database. The Ministry of Construction and Architecture maintains a Unified State Register of Highways that contains all the data collected for the state and local roads. The inventory data collected by the administrations of the five regions and Ashgabat are entered into this database.

Data analysis. Network data analysis is not currently carried out in Turkmenistan. Instead, annual planning is carried out based on the visual assessments by the road enterprises. These assessments form the basis for the annual work plans.

Performance-Based Road Maintenance Contracts

Performance-based contracting is not currently implemented in Turkmenistan. Road maintenance, repair, and construction works are carried out in-house through the state-owned road enterprises. This involves input-based payments where the costs of the different inputs are covered under the annual budget allocations.

Way Forward

Road asset management system. The government currently does not have a comprehensive data set for the road network that allows it to define the existing inventory, determine the current and future treatment needs, or carry out an objective prioritization of these needs. It is recommended that the government develop a basic RAMS with limited data collection for the road network as a first step. This may include basic data related to road location (GPS), length, roughness, and surface distress, allowing an analysis to be carried out of treatment and budget needs as well as treatment prioritization.

Performance-based road maintenance contracts. Turkmenistan has not expressed a strong interest in increasing the involvement of the private sector in road repair and maintenance, or in the introduction of performance-based contracting with the private sector. However, as is the case with other countries in the CAREC region, there is a potential for introducing elements of performance-based contracting in the maintenance and current repair activities carried out by the state-owned force account units through service-level agreements (SLA). This would introduce a shift from input-based to performance-based implementation, facilitating annual budgeting and introducing a degree of commercialization to the road enterprises. In the new toll roads that are being constructed through domestic companies, there is potential to introduce performance standards in the subsequent maintenance of these roads, with financing provided through the collected toll revenue. A new PPP Law issued in 2021 will likely lead to an increase in private sector involvement and the need for defining suitable performance standards for use in these contracts.

Uzbekistan

Road network. Uzbekistan has a total 141,882 km of registered roads, of which 42,869 km are classified as public roads. These include 3,993 km of international roads, 14,203 km of national roads, and 24,673 km of local roads. Resolution 226 of 2006 on Improving the Organization and Ensuring Quality Control of the Construction and Operation of Highways distinguishes between maintenance (excluding pavement and structure repairs), current repair (pavement and structure repairs), midterm repair (periodic maintenance), capital repair (rehabilitation), and (re)construction. In 2011, the resolution was updated, and midterm repair was merged with current repair. Current repair is currently defined to include both routine pavement and structure repairs as well as partial renewal of the wearing course. In practice, however, current repair tends to focus on annual pavement and structure repairs.

Institutional framework. The newly established Ministry of Transport (MOT) through its Committee for Roads is responsible for the management of the public road network. The Committee for Roads was created through the merger of the former joint stock company Uzavtoyul and the Republican Road Fund (RRF). In 2019, the road construction enterprises under the Committee for Roads were transferred to joint stock company Yulkurilish under the Agency for Management of State Assets and are currently being corporatized as state-owned companies. All new (re)construction and capital repair (rehabilitation) works are tendered competitively, with procurement carried out by the state unitary enterprise Directorate for the Construction and Reconstruction of Public Roads under the Committee for Roads. Current repair and maintenance continue to be carried out by regional and district-level road maintenance enterprises under the Committee for Roads that are contracted directly without competition. Since 2019, these enterprises are fully financed through the contract amounts awarded to them (including staff costs), with contracting managed by the regional road departments under the Committee for Roads. A number of design and research institutes and enterprises are also under the Committee for Roads. The former RRF was converted into Avtoyulinvest under the Committee for Roads in 2019 and currently focuses on donor-funded projects, while a newly created Republican Trust Fund for Road Development under the Ministry of Finance is responsible for financing (re)construction, repair, and maintenance of public roads.

Financing. Public roads used to get funding from the RRF, which received revenue from road user charges complemented by external loans. With the newly created Republican Trust Fund for Road Development, funding for (re)construction and capital repair of public roads comes from annual allocations from the state budget and interest-free loans from the Fund for Reconstruction and Development, while current repair and maintenance is financed from the budgets of the Republic of Karakalpakstan, the 12 regions, and the city of Tashkent. Although financing has increased over time in Uzbekistan sum terms, the devaluation of the sum has resulted in a decrease in financing in US dollar terms. The RRF budget reduced from $950 million in 2014 to just under $600 million in 2018. The portion of the budget executed through Uzavtoyul and later the Committee for Roads has varied between $450 million and $550 million per year. Expenditure on current repair and maintenance was steady at $205 million per year between 2016 and 2018, but subsequently decreased after the abolishment of the RRF to $145 million in 2019 and $130 million in 2020. The decrease in funding was especially evident in current repair, where the expenditure was halved. An HDM4 strategy analysis carried out in 2014 estimated that approximately $430 million per year is required to cover the annual costs of repair and maintenance (excluding the costs of backlog maintenance), or triple the current expenditure.

Road user charges. The RRF was financed largely from road user charges (75%–90% of all financing between 2013 and 2019). Most of the funding came from a corporate turnover tax that provided 55%–70% of all RRF financing ($335 million in 2018), but this was abolished in 2019 together with the RRF. A foreign vehicle entry fee provided approximately $21 million to the RRF in 2018, but is now allocated to covering the operational costs of the Committee for Roads and its underlying departments and management enterprises. Vehicle purchase fees provided approximately $133 million to the RRF in 2018, but now go directly to the state budget. Other important road user charges collected by the government include the fuel production excise tax ($190 million in 2018), the fuel consumption excise tax ($185 million in 2018), and a vehicle excise tax ($200 million in 2018). These were not earmarked to the RRF and are allocated directly to the state budget or to the regional budgets. Several other smaller road user charges were introduced in 2019 (regional fuel levy, parking fees, vehicle inspections, roadside advertising), the revenue of which is allocated to local road funds for financing urban and rural roads or to the operational costs of the Committee for Roads and its management enterprises. The introduction of tolling is currently being considered for financing new PPP expressway projects. The revenue currently collected from road user charges is more than sufficient to cover the estimated current repair and maintenance costs for public roads. However, the abolishment of the RRF has led to the loss of earmarked road user charges for public road maintenance and repair, resulting in a dependence on annual budget allocations by the regional governments (that are not responsible for public roads) and a significant decrease in financing and coverage.

Road Asset Management System

Development. ADB supported the development of a RAMS in 2010 under the CAREC Regional Road Project and again in 2014 under the CAREC Corridor 2 Road Investment Program. In both cases, consultants collected data for a portion of the public road network which was entered into a basic database and used to carry out an HDM4 analysis. However, these efforts did not lead to a sustainable RAMS or related capacities. This was further complicated by the involvement of both Uzavtoyul and the RRF in RAMS development, and disagreement regarding the responsibility for operation and management of the RAMS. The ongoing ADB-funded CAREC Corridor 2 Road Investment Program includes a RAMS expert position to provide direct support to the Committee for Roads in the development of its RAMS. A recent ADB technical assistance for preparing a Road Subsector Development Strategy and Action Plan included support to the development of a draft RAMS Action Program. Extensive support is also planned by the World Bank under the Regional Road Development Project, including the procurement of data collection equipment and a RAMS consultant to support data collection, development of a RAMS database, development of data analysis modules, and training of the Committee for Roads staff in the operation of the RAMS.

Data collection equipment. Much of the data collection equipment that was to be procured with ADB support was canceled due to complications with the government's procurement procedures. In part, this was found to be due to a disconnect between the data collection procedures described in existing standards and the proposed equipment for RAMS data collection. In the end, only a Roughometer was procured, linked to an accelerometer attached to the rear axle of a vehicle. This lacked a GPS module and was only linked to the vehicle odometer.

Under the Regional Road Development Project, procurement has started for an extensive set of network survey equipment. This includes a pavement condition survey vehicle that can record data regarding chainage, GPS, IRI (left and right wheel path), transverse profile, potholes, cracks, forward-looking video, downward-looking pavement video, event logging, longitudinal slope, crossfall, and curve radius from a single drive-over survey. Also to be provided with the survey vehicle are two pickups and three trailers with FWD, skid resistance-measuring equipment, and ground-penetrating radar. The project also seeks to procure 10 fixed automatic traffic counters (piezo-electric) and two mobile traffic counters (pneumatic tubes) capable of distinguishing up to 12 vehicle categories. To address the data needs for preparing road passports, the government has included the procurement of a road scanner and related vehicle. This involves a 360-degree high-definition video camera system and a LiDAR system with laser scanners to record all elements within the right-of-way and link it to GPS coordinates. The data collected with this equipment requires significant post-processing, increasing the data collection costs. The equipment will be provided to the state unitary enterprise Design and Survey Institute Uzyulloyikha under the Committee for Roads, which will be responsible for subsequent data collection under the project with technical support from the project consultant.

Other road survey equipment is already available with the unitary enterprise, Yul Loyiha Expertise, under the Committee for Roads, which specializes in road diagnostics and preparing feasibility studies and designs for the Committee for Roads and its other subsidiary enterprises. It owns two Trassa mobile pavement survey vehicles provided by Rozdortech as well as other Rozdortech equipment for measuring skid resistance and pavement strength (FWD). Although this equipment is quite advanced, it is not clear how suitable it is for road network surveys, especially in terms of linking the measured data to GPS location and providing the data in a format that facilitates data transfer to a RAMS.[36]

Data collection. Data collection is currently limited to the spring and autumn surveys based on visual inspections. Decree UP-5890 of 2019 on Measures to Deeply Reform the Road System of the Republic of Uzbekistan aims to change this and sets a target for collecting a complete road inventory using new technologies. Under the ADB-funded CAREC Regional Road Project in 2010 and under the CAREC Corridor 2 Road Investment Program in 2014, data were collected using a Roughometer for approximately 4,000 km of public roads, focusing on international roads and important highways. Under the current World Bank-funded Regional Road Development Project, data collection is planned for all public roads (international, national, and local), but this will be limited to three out of 14 regions where the project is focused. This will make use of the equipment to be procured under the project. The equipment of Yul Loyiha Expertise does not appear to be used for network data collection, and is used mainly to verify the quality of newly constructed pavements. A full inventory and condition survey of the entire public road network is not yet available or planned.

Database. Decree UP-5890 of 2019 on Measures to Deeply Reform the Road System of the Republic of Uzbekistan sets a target for making the collected road inventory data available in an electronic database and linking this to a geographic information system (GIS). The databases developed under the ADB-funded CAREC Regional Road Project in 2010 and under the CAREC Corridor 2 Road Investment Program in 2014 involved simple MS Access databases. These were meant to store the data collected by the consultants and make it available for use in HDM4. They were not set up as permanent RAMS databases with extended functionality and lacked GIS mapping.

The current World Bank-funded Regional Road Development Project is supporting the development of a simple RAMS database that will be linked to GIS mapping and will make use of freeware or commonly available software. This is to contain data collected by the road survey vehicle as well as data from the spring and autumn surveys. The database is to be provided to the three project regions, together with the necessary computer hardware and software. It appears that the database will not be set up as a central database within the Committee for Roads with remote access from the regions, but rather as a set of decentralized databases at regional level. This may complicate the planning of road interventions, especially for international and national roads that pass through different regions.

---

[36] Similar vehicles in Tajikistan and the Kyrgyz Republic have presented difficulties in this regard.

Data analysis. Decree UP-5890 of 2019 on Measures to Deeply Reform the Road System of the Republic of Uzbekistan sets a target to introduce new technologies for the selection and assessment of priority road projects. It further mentions the need to forecast the effect of these priority road projects on future road network conditions. This requires the analysis of collected data to determine repair and maintenance needs, estimate costs, and predict the impact of proposed works on future road conditions. Under the ADB-funded CAREC Regional Road Project in 2010 and the CAREC Corridor 2 Road Investment Program in 2014, data analysis was carried out using HDM4. Despite the fact that training was provided to staff, the government found HDM4 to be complicated to use, in part due to the lack of a Russian language interface and the high data requirements. A planned rollout of the HDM4 software to the regional road departments under the Committee for Roads was canceled.

The software to be used under the World Bank-funded Regional Road Development Project has not yet been defined, although HDM4 is mentioned together with other software such as RONET. Routine maintenance and current repair programs are to be based on the spring and autumn surveys, while a periodic pavement maintenance program will be prepared based on the data collected by the road network survey vehicle using a simple decision model. Based on the data analysis and planning tools to be used, the project consultants are to prepare annual and 5-year maintenance and rehabilitation plans.

Performance-Based Road Maintenance Contracts

Pilots. PBCs have not yet been piloted in Uzbekistan. ADB has been providing support to the government to introduce PBCs. As part of this support, the CAREC Corridor 2 Road Investment Program developed draft bidding documents for different types of PBCs covering 6 years of current repair and maintenance. This included a traditional output- and performance-based road contract (OPRC) to be applied with contractors, involving both rehabilitation and subsequent current repair and maintenance. However, because rehabilitation and maintenance were managed by different entities at the time (by RRF and Uzavtoyul, respectively) and because all current repair and maintenance was carried out by state-owned road maintenance enterprises, a second option was also proposed that consisted of performance-based SLAs for current repair and maintenance contracts to be applied with the state-owned enterprises. However, neither option was ever piloted.

Under the ongoing Kashkadarya Regional Road Project, a maintenance manual will be prepared for use by the road maintenance enterprises under the Committee for Roads, which will include performance standards, procedures, and contractual documents for the execution of performance-based road repair and maintenance works. Although Uzavtoyul and the RRF have now been merged into the Committee for Roads, current repair and maintenance continue to be contracted directly to the Committee for Roads' enterprises while other works are competitively tendered, complicating the implementation of OPRCs. The ADB technical assistance for preparing a Road Subsector Development Strategy and Action Plan has identified a need to significantly increase the volume of current repair (mainly periodic maintenance) of public roads, and has suggested that the required capacity increase may be partially provided by the private sector. It also suggests that this may form an opportunity for piloting OPRC-type contracts with private sector contractors, including a combination of output-based periodic maintenance with several years of performance-based routine maintenance.

Performance standards. As part of the draft bidding documents prepared by the CAREC Corridor 2 Road Investment Program, a set of performance standards were also prepared. The performance standards are similar for the different contract types and apply to current repair and routine maintenance activities, but do not include winter maintenance. Some of the performance standards are not measurable, complicating inspections and potentially leading to disputes regarding compliance. The compliance with the performance standards is verified through monthly formal inspections carried out in a sample of the contracted road length.

Compliance is checked for a group of performance standards rather than for each individual standard, and takes account of the number of instances of noncompliance. Contractors are given time to correct identified defects and penalties are applied only in case the defects are not corrected within the stipulated response times. Response times are provided even where the defect involved can easily be predicted and avoided (e.g., vegetation height), undermining the nature of performance-based contracting. The calculation of the penalty amounts and the resulting deduction to the monthly lump-sum payments is not clearly defined in the bidding documents and needs to be improved. Further review of the performance standards and related response times and penalties is required.

Way Forward

Road asset management system. RAMS development in Uzbekistan has received intermittent support over the years, but has not yet led to a comprehensive or sustainable system. A coordinated medium-term approach is required to ensure that the different elements of RAMS development are addressed, and that sufficient support is provided to institutionalize the required skills and procedures. ADB has supported the Committee for Roads in the preparation of a RAMS Action Program that lays out the required steps in RAMS development and institutionalization over the period 2021–2025, distinguishing steps in data collection, database development, and data analysis, as well as the encompassing institutional framework and sector financing. Once issued by the Government of Uzbekistan, the RAMS Action Program will allow for greater coordination between development partners in support of its implementation, building on ongoing and future support efforts to develop and institutionalize the RAMS. Major components of the RAMS development will involve the collection of inventory, condition, and traffic data for the full public road network, the development of a centralized RAMS database with remote access, and the preparation of a suitable decision model for identifying and prioritizing treatment needs. The establishment of a dedicated RAMS unit under the Committee for Roads responsible for operating the RAMS, complemented by annual budget allocations for data collection and operation, is a further requirement for the success of RAMS development in Uzbekistan.

Performance-based road maintenance contracts. Although development partners have supported the introduction of PBCs, the approach has not yet been piloted in Uzbekistan. With all current repair and maintenance carried out through direct contracts with road maintenance enterprises under the Committee for Roads, this provides an opportunity for introducing performance-based SLAs between the Committee for Roads and these enterprises, providing clear benefits over the current input-based contracts. This may initially be piloted on a small scale with a limited number of enterprises, and later replicated with other enterprises, if successful. Given the need to significantly increase the volume of periodic maintenance in the public road network and the expected involvement of the private sector to provide the required capacity, this forms an opportunity to also pilot OPRCs. Both options will require a detailed analysis of existing legislation and standards that may form an obstacle for applying such contracts. Based on the results of these pilots, a decision may subsequently be taken by the Committee for Roads on how to continue with the contracting of repairs and maintenance in the future.

Road Asset Management System

Status. RAMS development in the CAREC region has achieved significant progress in the last 5–10 years. Almost all countries have collected some road network data, have a database to store and access the data, and have carried out some type of analysis of the collected data to support planning and monitoring. From the initial **piloting phase** involving data collection in a small part of the road network using basic equipment, and development of a basic database for data storage and data analysis carried out by consultants, RAMS development in most countries has progressed toward the **replication phase**, where expanded network-level data are collected using network survey vehicles, comprehensive databases are developed with extensive road management functionality, and specific planning modules or software are used in support of data analysis and planning. This is often complemented by the establishment of a dedicated RAMS unit and the creation of in-country capacity for replicating the data collection, database management, and data analysis. Despite

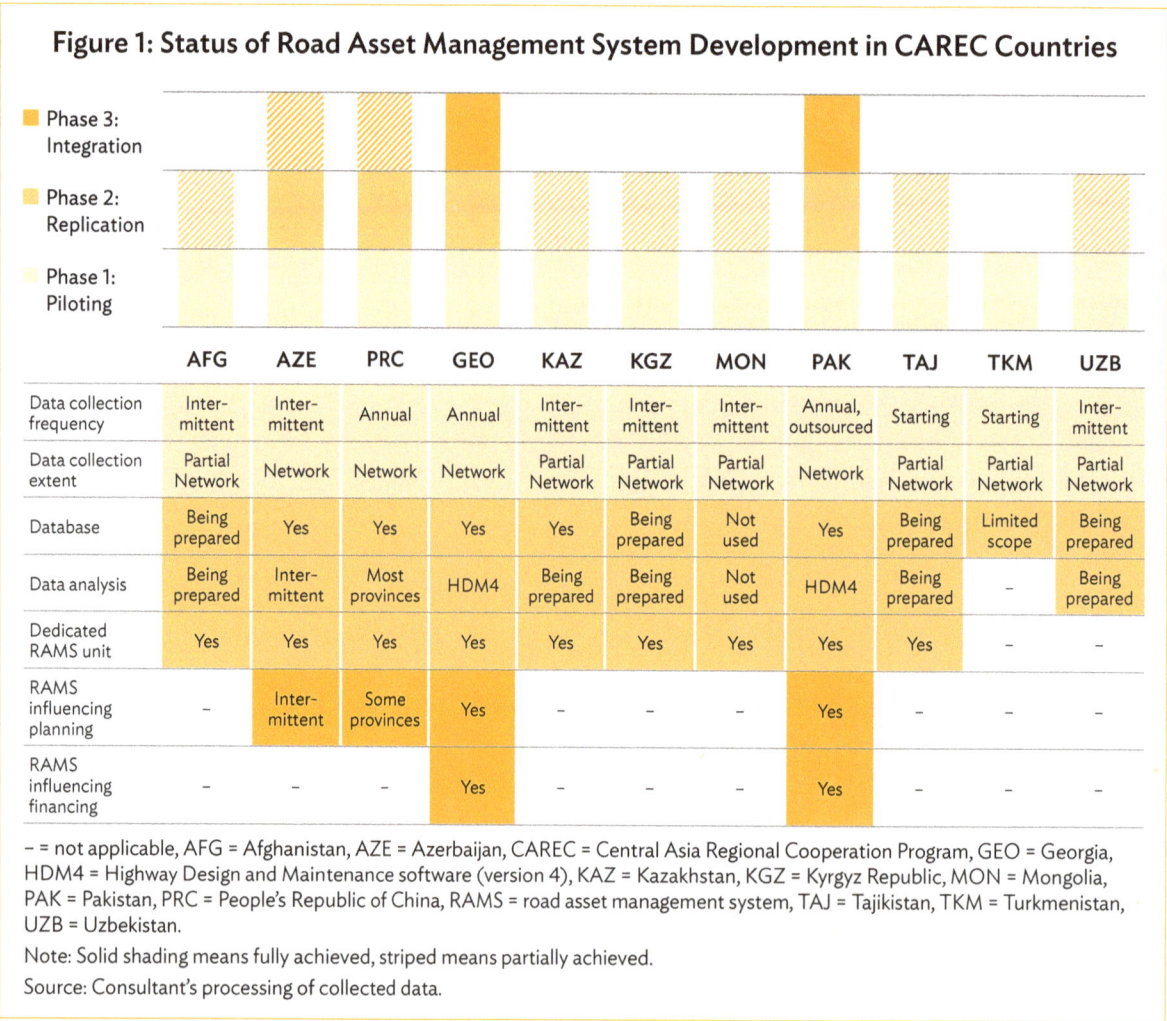

Figure 1: Status of Road Asset Management System Development in CAREC Countries

	AFG	AZE	PRC	GEO	KAZ	KGZ	MON	PAK	TAJ	TKM	UZB
Data collection frequency	Inter-mittent	Inter-mittent	Annual	Annual	Inter-mittent	Inter-mittent	Inter-mittent	Annual, outsourced	Starting	Starting	Inter-mittent
Data collection extent	Partial Network	Network	Network	Network	Partial Network	Partial Network	Partial Network	Network	Partial Network	Partial Network	Partial Network
Database	Being prepared	Yes	Yes	Yes	Yes	Being prepared	Not used	Yes	Being prepared	Limited scope	Being prepared
Data analysis	Being prepared	Inter-mittent	Most provinces	HDM4	Being prepared	Being prepared	Not used	HDM4	Being prepared	–	Being prepared
Dedicated RAMS unit	Yes	Yes	Yes	Yes	Yes	Yes	Yes	Yes	Yes	–	–
RAMS influencing planning	–	Inter-mittent	Some provinces	Yes	–	–	–	Yes	–	–	–
RAMS influencing financing	–	–	–	Yes	–	–	–	Yes	–	–	–

– = not applicable, AFG = Afghanistan, AZE = Azerbaijan, CAREC = Central Asia Regional Cooperation Program, GEO = Georgia, HDM4 = Highway Design and Maintenance software (version 4), KAZ = Kazakhstan, KGZ = Kyrgyz Republic, MON = Mongolia, PAK = Pakistan, PRC = People's Republic of China, RAMS = road asset management system, TAJ = Tajikistan, TKM = Turkmenistan, UZB = Uzbekistan.

Note: Solid shading means fully achieved, striped means partially achieved.

Source: Consultant's processing of collected data.

the significant progress, only a couple of countries have reached the **integration phase**, where the RAMS is integrated into existing planning and budgeting procedures and influences financing levels and budget allocations. Georgia and Pakistan have achieved this level of RAMS development, although this does not mean the end of RAMS development has been reached. Both countries are currently expanding their data collection and further improving the functionality of their databases, to strengthen data analysis and form a better basis for planning and budgeting. Both countries are also moving from in-house operation of all RAMS activities toward the outsourcing of these activities and also some planning activities.

Complexity. Of particular interest in Georgia and Pakistan is the fact that their initial RAMS was relatively simple, involving limited data collection, straightforward databases, and off-the-shelf software for data analysis and planning. Only now that their RAMS has been fully integrated into existing systems and procedures and they have identified additional needs, are they working on expanding the data collection and the database functionality. This should be compared to other CAREC countries that have rapidly moved to complex systems with extensive data collection, comprehensive database systems, and complicated custom-made planning modules. The costs involved in the regular data collection together with a lack of acceptance of the results of the custom-made planning modules, often forms an obstacle to the continued use of the RAMS in these countries. This issue is already identified in ADB's *Compendium of Best Practices in Road Asset Management* (footnote 2), which recommends limiting the data to be collected, keeping the database simple, and using easy-to-use planning modules or off-the-shelf software. These recommendations continue to be valid. CAREC countries should therefore initially limit their data collection, in the understanding that this can be expanded at a later stage. Where custom-made software is applied, this should initially use a simple database structure and tested approaches, such as decision matrices, based on detailed analyses of collected data, rather than developing comprehensive systems from scratch. To the extent possible, local software developers should be involved to facilitate future support.

Network coverage. There appears to be a misunderstanding in many CAREC countries that the data collection for a RAMS necessarily needs to cover the entire road network. Although the RAMS cannot be used to plan individual interventions in roads for which data has not been collected, it can be used to plan interventions for a subnetwork for which data are available, and it may also be used to estimate overall budget needs for the subnetwork for which not all data has been collected. Especially in countries where a large length of local roads is centrally managed, data collection for the entire network can be quite expensive. The budgets in these countries are often insufficient to cover all road maintenance and repair needs, with a major portion of the budget allocated to higher-level roads and only limited allocations to local roads. In such cases, it may be preferable to initially focus the RAMS on the subnetworks of higher-class roads which are more likely to receive funding to carry out the identified treatment needs. Although such a distinction can be made based on road class, it is more appropriate to make it based on surface type and traffic level. A RAMS will generally prioritize budget allocations to the repair and maintenance of roads that have higher traffic volumes and higher pavement standards. Where available repair and maintenance budgets are limited, the RAMS will allocate most of the budget to these important roads. It therefore makes sense to initially focus the data collection on these important roads, expanding the data collection to include less important roads only at a later stage when more financing becomes available to address the treatment needs in these roads. This may include the identification of a core road network that includes the most important roads from the point of view of traffic levels, but also in terms of providing access to neighboring countries, different parts of the country, or administrative centers. Definition of such a core network allows data collection and analysis as well as subsequent treatments to be focused on these roads, ensuring a minimum access standard and service level is achieved. This may then at a later stage be expanded to include a greater portion of the entire road network.

Continued support. The successful development of a RAMS takes time. This necessarily needs to go through the different stages of piloting, replication, and integration, with the duration of each stage progressively becoming longer. Where piloting may be completed in a couple of years, the RAMS replication phase can easily take 5 years or more, and together with the integration of the RAMS will easily take 10 years or more. As the RAMS becomes integrated and is used regularly, additional needs and functionalities will often be identified that require further development (collecting additional data types, using more advanced equipment to facilitate data collection, increasing the functionality of the database to respond to needs, expanding the RAMS to include a bridge/tunnel management system or a contract management system, further improving the planning module to better reflect needs and priorities, etc.). Georgia started development of its RAMS in 2008, and is currently planning to expand the data collection and increase the functionality of its database. Pakistan has been developing its RAMS since 1999, and is now moving toward the outsourcing of data collection and data analysis. The successful introduction of a RAMS requires continued support from the government and from development partners over an extended period. This cannot be achieved in a single project. It will require support from successive projects and possibly from different development partners to ensure that the RAMS becomes fully operational. With many of the CAREC countries currently in the RAMS replication phase, it is important that they ensure they receive continued support to advance RAMS development and integration.

Integration. With most CAREC countries reaching the replication phase and expected to have an operational RAMS in the coming years, there is an urgent need to focus on the next phase of integrating the RAMS into existing systems and procedures. This includes institutionalization of the RAMS operation, integration of the RAMS into existing planning and budgeting procedures, allocation of appropriate financing levels to respond

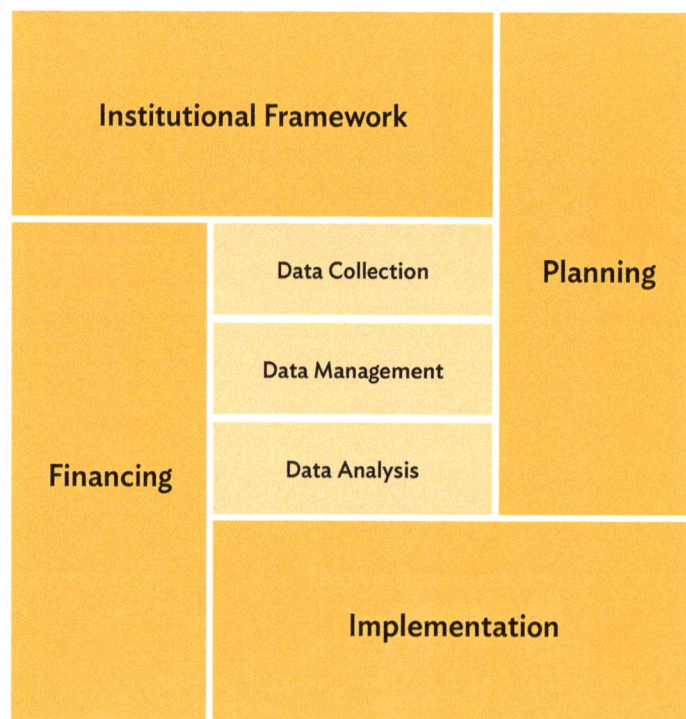

Figure 2: Road Asset Management System Integration

Institutional Framework

Financing

Data Collection

Data Management

Data Analysis

Planning

Implementation

Source: ADB. 2018. *Compendium of Best Practices in Road Asset Management.* https://www.adb.org/sites/default/files/publication/396126/best-practices-road-asset-management-carec.pdf.

to the needs identified with the RAMS, and development of the implementation capacity to carry out the prioritized works. This is reflected in the figure below, showing how the RAMS fits into the wider framework of road management. The integration phase should take place in parallel with the replication phase.

Institutionalization. With most CAREC countries having reached the replication phase, the institutionalization of the RAMS needs to receive particular attention to ensure that RAMS operation becomes sustainable. Many countries have established a dedicated RAMS unit and have involved its staff in data collection, database management, and data analysis, generally with support from consultants. The identification of dedicated staff and the development of their capacities is a first requirement to ensure sustainability of the RAMS and future replication of RAMS activities by the countries concerned. This alone is not enough, however, and an important obstacle faced in many countries is the lack of sufficient budget and resources to carry out RAMS activities on a regular basis. Data collection is often especially affected by a lack of funding, resulting in outdated data that does not allow for proper planning. Data need to be collected on a regular basis to keep the RAMS functional. Generally, this involves collecting data for a portion of the road network each year, avoiding large peaks in data collection and related budget requirements. This may be done in-house by government staff where networks are small and data needs are limited, but will increasingly need to be outsourced, providing certain survey equipment to the consultants where this is required. The operation of the RAMS database and the data analysis will also require some budget to cover the costs of preparing reports and maps, internet access and server storage, and replacement or upgrading of survey and computer equipment every few years. Providing adequate annual budget allocations for RAMS operations can help ensure the sustainability of the RAMS. This may be combined with a requirement for detailed annual reporting on the road network based on the collected data, providing a clear incentive for providing the necessary staffing and budget in order to prepare the required reports (as has been the case in the PRC).

Planning and budgeting procedures. A subsequent step is the integration of the RAMS into existing planning and budgeting procedures, ensuring that RAMS analysis forms the basis for annual and multi-annual planning and budgeting. Even in the case of a fully operational RAMS, this will have little benefit if it is not able to influence annual planning and budgeting. Integration into existing planning procedures should be given attention at an early stage, ensuring that the procedures for data collection, database management, and data analysis fit within the time schedule and needs of annual planning and budgeting procedures. This is not always straightforward, as many CAREC countries continue to rely on planning procedures where needs are identified at the local level, often by in-house maintenance and repair units. The identified needs are subsequently consolidated at the regional level and again at the central level. This has the disadvantage that priorities are largely defined at the local level based on subjective criteria, and that the list of treatment needs received at the central level reflects only a portion of the full needs. This is very different from the approach used in the case of a RAMS, where the identification of needs and the planning and budgeting is carried out at the central level, based on data collected for the entire (sub) network of roads. Although this depends on data collected at the local level and draft plans are often shared with local units to confirm identified needs and priorities, it involves a fundamental shift in the procedures applied in planning and budgeting. This often requires adjustments to legislation and regulations to allow the RAMS and related data analysis to form the basis of planning and budgeting, and particular attention should be given to address any legal obstacles on using the RAMS for annual planning (the Kyrgyz Republic, for instance, is well on its way in achieving this). Significant capacity building will also be required to make staff familiar with the new procedures. The final plans and budgets will not be fully determined by the RAMS, and further adjustments will need to be made to the treatment needs and priorities identified by the RAMS. This will take into account other criteria (e.g., accessibility, contract packaging), in addition to the criteria used in the RAMS. The procedures for the subsequent steps after the RAMS analysis will also need to be clearly defined. The integration of the RAMS into the planning and budgeting procedures is likely the most difficult step in RAMS development and will require clear government support.

Appropriate funding for road maintenance and repairs. A third important area in the integration phase is to ensure appropriate funding is available for financing the plans developed with the help of the RAMS, especially regarding road maintenance and repairs. Even the most perfect plan will be useless if sufficient funding is not available to implement it. Using the RAMS, a proper assessment can be made of the road maintenance and repair needs and related funding requirements for the entire network. This should be linked to the identification of suitable financing sources to cover the identified funding requirements. International experience shows that road user charges are very suitable for financing road repair and maintenance, introducing a user-pays principle. Revenue from road user charges is generally linked to the number of vehicles and the degree of road usage, allowing accelerated road deterioration from increased usage to be compensated by increased revenues from the road user charges. To improve the level and predictability of road maintenance funding, such road user charge revenues should be earmarked for the road sector, providing predictable funding levels that facilitate multi-annual planning and contracting. This may be complemented by the creation of a specific road maintenance account or road maintenance fund to support the management of the funding. In doing so, it is important that such earmarked road user charge revenues prioritize funding use for maintenance and repairs, ensuring that existing roads are properly maintained, and that road deterioration is slowed down.

Works implementation capacity. A final focus area in the integration phase is the capacity for implementing the planned maintenance and repair activities. Where a RAMS is successfully used to ensure increased funding is made available for road repair and maintenance, this will lead to greater volumes of works to be carried out each year. The use of a RAMS and the related prioritization based on economic benefits also tend to result in a shift toward the maintenance and repair of roads in good and fair condition, avoiding that these roads deteriorate to poor condition and require costly rehabilitation. Especially in the CAREC countries where the implementation of periodic maintenance (midterm repair) has been limited in the past, this will likely lead to a significant increase in the volume of periodic maintenance to be carried out (as evidenced in Georgia and Pakistan). This will require strengthening of the implementation capacity in the different countries. Although the required capacities may be created in state-owned enterprises that are involved in maintenance implementation in many CAREC countries, this forms a unique opportunity to promote the participation of private sector contractors without immediately affecting the existing road maintenance enterprises. Where appropriate, this may involve the introduction of performance-based contracts to cover periodic maintenance as well as subsequent routine maintenance.

Action program. Each CAREC member country is different, having a different level of RAMS development, different past experiences, different systems and procedures, different revenues and budget levels, different road networks, etc. This document provides an overview of the different countries and the status of RAMS development in each of them, highlighting some of the specific challenges and achievements in each country. However, this document does not achieve the level of detail that is required to determine the specific actions to be undertaken in each country in its development of RAMS. The general recommendations provided in this document need to be worked out in further detail in the form of a RAMS Action Program. This should be done in close collaboration with the authorities responsible for road management and with the involvement of the different development partners willing to support RAMS development. The RAMS Action Program should cover the next 5–10 years and should define the specific actions to be undertaken and how these will be financed. This should cover the basic elements of a RAMS (data collection, database management, and data analysis), but should also be expanded to include the institutionalization and integration of the RAMS, linking it to institutional frameworks, planning and budgeting procedures, road sector financing, and implementation modalities. Such RAMS Action Programs should be prepared for each CAREC member country, guiding future support to RAMS development.

Performance-Based Road Maintenance Contracts

Status. Development of PBCs in the CAREC region has not progressed as much as that of RAMS. **Piloting** of PBCs has been carried out in the majority of CAREC countries. Some countries have only developed performance standards, but have yet to apply these in pilot contracts. Several countries have progressed beyond this initial piloting phase and are moving toward the **replication** phase, implementing subsequent PBCs and incorporating the lessons learned from previous pilots. Although in some cases the funding has come from government, all pilots have taken place under development partner projects. The third phase of **integration** into normal government contracting and mainstreaming of the performance-based contracting approach for road maintenance and repairs has not yet been reached by any of the CAREC countries.

Figure 3: Status of Performance-Based Road Maintenance Contract Development in CAREC Countries

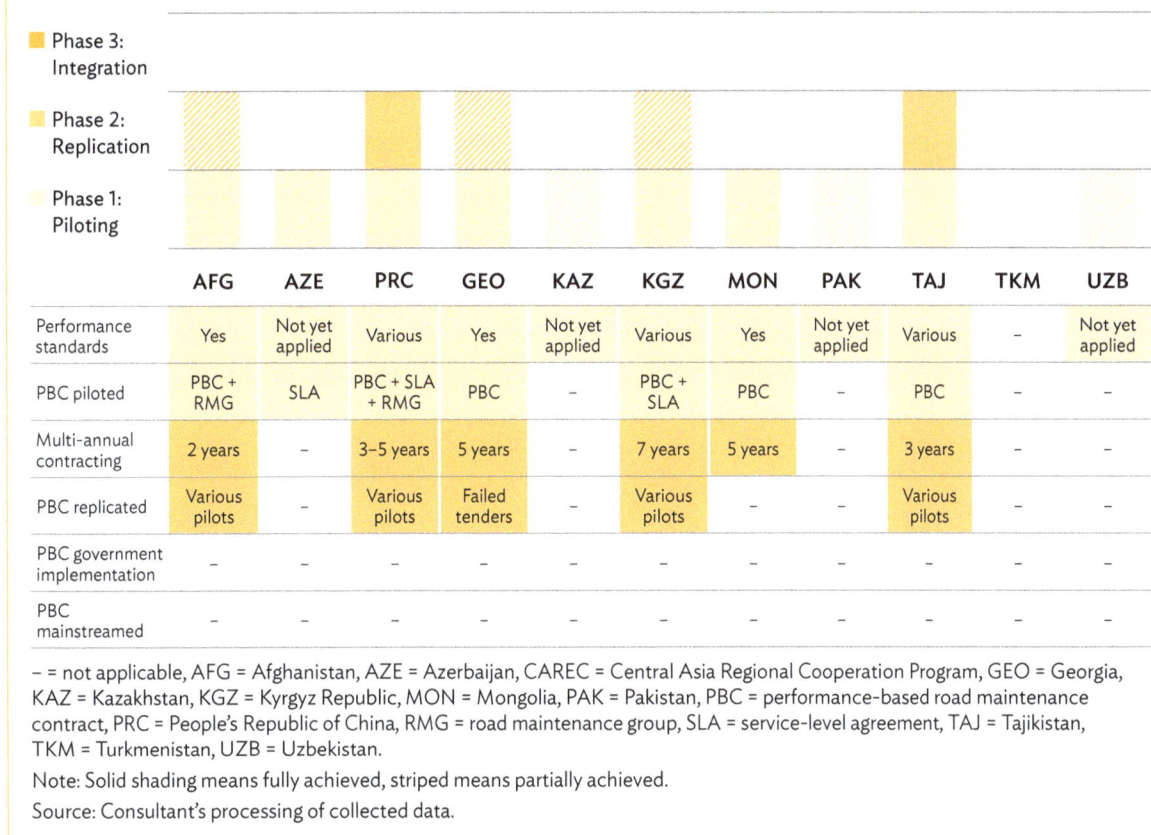

Phase 3: Integration

Phase 2: Replication

Phase 1: Piloting

	AFG	AZE	PRC	GEO	KAZ	KGZ	MON	PAK	TAJ	TKM	UZB
Performance standards	Yes	Not yet applied	Various	Yes	Not yet applied	Various	Yes	Not yet applied	Various	–	Not yet applied
PBC piloted	PBC + RMG	SLA	PBC + SLA + RMG	PBC	–	PBC + SLA	PBC	–	PBC	–	–
Multi-annual contracting	2 years	–	3–5 years	5 years	–	7 years	5 years	–	3 years	–	–
PBC replicated	Various pilots	–	Various pilots	Failed tenders	–	Various pilots	–	–	Various pilots	–	–
PBC government implementation	–	–	–	–	–	–	–	–	–	–	–
PBC mainstreamed	–	–	–	–	–	–	–	–	–	–	–

– = not applicable, AFG = Afghanistan, AZE = Azerbaijan, CAREC = Central Asia Regional Cooperation Program, GEO = Georgia, KAZ = Kazakhstan, KGZ = Kyrgyz Republic, MON = Mongolia, PAK = Pakistan, PBC = performance-based road maintenance contract, PRC = People's Republic of China, RMG = road maintenance group, SLA = service-level agreement, TAJ = Tajikistan, TKM = Turkmenistan, UZB = Uzbekistan.

Note: Solid shading means fully achieved, striped means partially achieved.

Source: Consultant's processing of collected data.

Lessons learned. With one or more PBC pilots carried out in the majority of CAREC countries, this forms an opportunity for a detailed review of these experiences to see what has worked well and what needs further improvement. This has not yet taken place, and there is a risk that the experiences gained in these PBC pilots will be lost and will not be considered in the design of future PBCs. Such a detailed review of PBC pilots may be carried out in an individual country, but may also involve several different CAREC countries, allowing lessons to be learned from other countries that are at the same stage of PBC development and have a similar framework within which the PBCs need to operate. This is especially true for the former Soviet Union countries that tend to have a similar institutional and legal framework, often with similar obstacles regarding the use of multi-annual contracts, the costing of maintenance activities, and the involvement of force account units. Such a detailed review should collect all details of the different pilots, allowing similarities and differences to be identified. It should include a detailed description of the performance standards used and the payment and deduction mechanisms applied, as well as an assessment of any problems faced in the use of performance standards and the application of deductions. But the review should also assess contract details related to the use of advances, performance guarantees, retention payments, defect liability periods, etc. The PBC pilots carried out in CAREC countries form a wealth of knowledge and experience that may help improve future PBCs, avoiding mistakes from being repeated, and allowing good practices to be replicated. Some initial data obtained in this review of the status of PBCs in different CAREC countries are reflected in the following table, showing the different types of contracts, contracted road lengths, contract durations, scope of works, and costs per kilometer per year in the various contracts.

Table: Overview of Performance-Based Road Maintenance Contracts in CAREC Countries

Country	Type	Length (km)	#	Duration (months)	Initial Repairs	Cost ($/km/year)	PBC Activities	PBC Cost ($/km/year)
AFG	PBC	142	1	36	–	2,800	RM+WM	2,800
AFG	PBC	1,626	5	26	–	6,300	RM+WM	6,300
AZE	SLA	774	3	12	–	15,000	RM+WM	15,000
PRC	PBC	57	1	60	Full RH+PM	33,250	RM	
PRC	SLA	107	1	36	Partial RH+PM	25,000	RM	
PRC	PBC	120	2	24	Partial RH+PM	20,000	RM	
PRC	PBC	776	8	60	Partial RH+PM	29,000	RM	
GEO	PBC	117	1	60	Partial RH	28,500	PM+RM+WM	5,200
GEO	PBC	240	1	60	Partial RH+PM	Canceled	RM+WM	
GEO	PBC	140	1	60	Limited RH	Planned	RM+WM	
GEO	PBC	150	1	60	Partial RH+PM	Planned	RM+WM	
KAZ	PBC	1,415	4	96	Partial RH+PM	Canceled	RM+WM	
KGZ	SLA		1	12	–	N/A	RM+WM	
KGZ	PBC	69	1	36	Limited RH+PM	21,000	RM+WM	1,500
KGZ	PBC	70	2	84	Full RH	82,000	RM+WM	8,700
MON	PBC	58	1	60	Full RH	49,000	RM+WM	
PAK	PBC	104	2	60	Limited RH+PM	20,000	RM+WM	
TAJ	PBC	149	2	36	Limited RH+PM	7,600	RM+WM[a]	5,800
TAJ	PBC	176	2	36	–	1,500	RM+WM[a]	1,500

– = not applicable, AFG = Afghanistan, AZE = Azerbaijan, GEO = Georgia, KAZ = Kazakhstan, KGZ = Kyrgyz Republic, km = kilometer, MON = Mongolia, PAK = Pakistan, PBC = performance-based road maintenance contract, PM = periodic maintenance (midterm repair), PRC = People's Republic of China, RH = rehabilitation (capital repair), RM = routine maintenance (including current repairs), SLA = service-level agreement, TAJ = Tajikistan, TKM = Turkmenistan, UZB = Uzbekistan, WM = winter maintenance.

[a] Winter maintenance was paid separately on an output basis.

Source: Consultant's processing of collected data.

Costs. The table shows significant differences in the costs per kilometer per year of the different PBCs. This depends strongly on the scope of works included under the contract, primarily on the scope and length of initial repairs included under the contract. But even when only the performance-based component of the contract is assessed (generally covering routine maintenance and current repairs as well as winter maintenance), significant differences are visible between countries and even within countries. To a certain extent, this reflects differences in road condition, traffic, topography, climate, etc., but these cost differences also reflect the level of risk perceived by the contractors. A proper distribution of risks between client and contractor can significantly reduce the perceived risks and related bid prices. As an example, after the initial pilots in Georgia, it was decided to change from performance-based to output-based payments for periodic maintenance, while in Tajikistan, the decision was taken to pay for the unpredictable winter maintenance on an output basis with only routine maintenance and repair paid on a performance basis. We also expect to see a gradual reduction of costs in subsequent PBC pilots in the same country as contractors gain experience with the performance-based contracting approach and are better able to estimate the risks involved. As experience is gained, competition also tends to increase, further reducing costs. In those countries where some analysis of the costs has been carried out, the PBC costs were found to be more or less in line with traditional maintenance costs, while often achieving a higher standard. A further analysis of the costs of PBCs in the different countries, distinguishing the different activities involved and identifying the possible reasons for cost differences, would be very useful in designing future PBCs.

Contract duration. The contract duration is of particular importance in PBCs. An appropriate contract duration seeks a balance between the foreseeable maintenance needs, the returns on investments in equipment and materials, and the risk of unforeseen maintenance needs. Short contract durations include relatively little foreseeable maintenance (especially if they are preceded by rehabilitation or periodic maintenance works), but still need to cover the risks of unforeseen maintenance needs. The short duration and small contract amount also make it difficult to invest in equipment and purchase materials in bulk. As a result, bid prices tend to be relatively high in relation to the foreseeable maintenance needs. The use of such short contract durations may therefore result in costs that are higher than in the traditional output-based maintenance contracts, undermining the introduction of PBCs. Long contract durations, on the other hand, include more maintenance as the road deteriorates and becomes older, making it worthwhile to invest in equipment and purchase materials in bulk. However, the longer duration makes it harder to properly predict the maintenance needs and increases the risk of unforeseen maintenance needs, especially if the maintenance is not preceded by rehabilitation or periodic maintenance works carried out by the same contractor (of which the quality is known). As a result, the contractor will be forced to increase the bid price to cover these risks, again resulting in costs that are relatively high. If a proper balance is found, the risks of unforeseen maintenance needs can be kept low, while the extended contract duration allows the contractor to invest in equipment and procure materials in bulk, resulting in greater efficiency and lower costs. This optimal duration will depend on the type of contract. If the contract only includes routine (and winter) maintenance and repairs, the most suitable duration will be in the order of 3–5 years. Where output- and performance-based road contracts (OPRCs) are concerned involving initial rehabilitation or periodic maintenance, optimal contract durations will be in the order of 5–10 years. The indicated durations are the durations of the performance-based maintenance after completion of any initial works. The shorter contract durations may be applied initially, gradually extending the duration in subsequent contracts as contractors gain experience and are better able to assess the risks involved.

Performance standards. Performance standards form the heart of any PBC, defining the road condition targets to be achieved by the contractor to be eligible for the agreed lump-sum payment. Despite their importance, insufficient attention tends to be given to defining and testing the performance standards before the contract is signed, often leading to poor performance and the need for contract variations. In reviewing the performance standards applied in the different countries and PBCs, significant differences were visible in the types of indicators used to define the road condition, and in the threshold values applied. Even when comparing PBC pilots in the same country, performance standards tend to differ from one contract to the next. The performance standards often depend more on the consultants involved in their preparation, than on reviews of past experiences and identified improvement needs. Many of the performance standards continue to include weaknesses that can jeopardize the success of a PBC. These mainly relate to the definition of the performance indicator and the related thresholds. Many performance standards are not objectively measurable, use indicators that are not defined in a manner that facilitates inspection, use indicators that are not relevant to the activity to be carried out, or have thresholds that are set unrealistically low or high.[37] This can lead to unnecessarily high bid prices, unacceptable road conditions, and disputes regarding the compliance with performance standards. In all reviewed performance standards, there is room for improvement. A detailed review of the performance standards used in previous and ongoing pilots should be carried out, identifying which standards functioned well and which did not. This should be expanded to include various countries, providing a wider set of experiences from which the best options may be identified.

Response times. Performance standards are often linked to response times within which any noncompliance identified during the inspection has to be corrected. Payment deductions are only applied if the identified defects are not corrected within the allowed response times. Such response times are required for certain defects that cannot be properly predicted and that occur suddenly, such as landslides, heavy snowfall, flooding, traffic accidents, etc. However, for most road maintenance activities, the damages can be predicted and treated before the thresholds are exceeded. If some small potholes are formed, the contractor can plan to patch the potholes before the size or total number of potholes exceeds the allowable threshold. Vegetation can be cut before it exceeds the allowable length. For such defects that can be treated by the contractor before the threshold is reached, response times are not suitable and should not be used. Use of response times for such defects has in many countries caused contractors to await the results of the inspection, and only address those defects identified in the inspection. This undermines the nature of a PBC, where the contractor is expected to work proactively in avoiding and addressing defects. In addition, the use of response times requires additional inspection visits to check whether defects have been corrected within the allocated response time, increasing rather than reducing the burden of supervision and inspection. Where response times of different durations are used, this can result in the need for multiple inspection visits every month. In future PBC pilots, the response times should be avoided as much as possible, keeping these only for defects that cannot be predicted and that occur suddenly.

Payment deductions. Noncompliance with the performance standards in PBCs leads to deductions to the monthly lump-sum payments. The level of these deductions should vary by performance standard and should reflect the importance of the defect as well as the cost of repairing the defect (defects with high safety risks or that are costly to repair should have higher deductions). Where the deductions are very low, there is insufficient incentive for the contractor to carry out maintenance and repairs in a timely manner. The PBC pilots in CAREC countries include several cases with such low deductions. Very high deductions are also not suitable, as they form a financial risk for the contractor who will increase the bid price to reflect this risk. The deductions, therefore, need to be set at a level that reflects the costs of repairs and any risks created by the defect, ensuring that there is an incentive for the contractor to act in a timely manner without creating unnecessary financial risks.

Apart from the level of the deduction, the method of calculating and applying the deduction is also important. Common practice in PBCs is to divide the contracted road into 1 km sections, each linked to an equal portion of the agreed monthly lump-sum payment. The deductions for each performance standard can then be defined as a percentage of the monthly payment for that 1 km section of road, to be applied in case that section is found to be noncompliant. The total deduction in any given month is then the sum of the deductions applied to each of the 1 km sections. This results in a very transparent system and allows the same

---

[37] For instance, a maximum pothole size of 0.5 m² that is equivalent to an allowable pothole diameter of 80 cm, setting a maximum blockage of side drains of 50 meters instead of defining the maximum degree of blockage of the cross section, or not allowing any sedimentation in culverts which will be very costly to ensure.

approach to be used in different roads and contracts, irrespective of the road length or bid price. This can be expanded to allow full payment in case of limited noncompliance, to include bonuses in case of extended periods of full compliance, or to avoid deductions in the first months of operation. The important thing is that the system of calculating the deductions is kept simple and transparent, so that it is clear for the contractor what the impact will be if certain performance standards are not complied with. This allows the contractor to properly assess the financial risks of potential deductions, and calculate the costs of providing sufficient resources to minimize these risks. Proper understanding of the performance standards and the deduction calculations has shown to lead to lower bid prices and to avoid disputes during contract implementation. Where performance standards are not clear, or related deduction calculations lack transparency, contractors will view this as a risk and reflect this in higher bid prices. Here, too, it is recommended that a proper review be carried out of the deduction systems applied in the different PBC pilots, reviewing how well these have worked and comparing these to the performance of the contractor and the resulting road conditions.

Integration. Integration refers to the application of the PBC approach under normal government contracting procedures, and the mainstreaming of the approach for road maintenance. To a certain extent, the lack of integration of the PBC approach in CAREC countries is due to existing legislation and funding policies that limit the scope for procuring and implementing PBCs. In some cases, this involves requirements that road maintenance and repairs are to be contracted on a volume basis, or regulations that define that the evaluation of bids should be based on the review of unit costs. In other cases, national legislation requires that payments be based on the volume of work completed. The application of multi-annual contracts can also form an obstacle, and even where such multi-annual contracts are allowed, it may not be possible to secure the required budget beyond the end of the financial year. Although in most cases it is possible to pilot PBCs in such circumstances under development partner procurement procedures, these legislative and funding issues need to be addressed if the aim of integrating the PBCs into normal government contracting for road maintenance is to be achieved. Currently some countries are opting to instead expand the application of PPPs incorporating performance-based aspects. Although PPPs are suitable in certain circumstances, they are generally more complicated than traditional PBCs and should not be seen as a direct alternative to amending existing regulations and standards related to traditional contracting of road maintenance and repair. A review of existing legislation and financing policies should, therefore, be included in the preparation of any future PBC pilots, or as part of the detailed review of past and ongoing PBC pilots. This should be combined with the development of concrete proposals identifying options for applying PBCs under government systems, including any necessary amendments of legislation or financing policies.

Action program. Although this document provides some recommendations regarding the actions to be undertaken in the further development of PBCs, these recommendations lack the detail required to convert them into specific actions that may be carried out by governments and supported by development partners. In order to support further development of PBCs in the different countries, a more detailed PBC Action Program should be prepared for each country, taking into account the specific characteristics of that country. This should be prepared in close collaboration with the authorities responsible for the road network, and with involvement of the development partners willing to support the further development of PBCs. The PBC Action Program should cover the next 5–10 years and should define the specific actions to be undertaken regarding the piloting of PBCs, the development of the performance standards, the use of response times, the application of payment deductions, the duration of the contracts, and the use of PBC bidding documents in each country. But the PBC Action Program should also cover the wider context of financing road repairs and maintenance, legal regulations and standards regarding road repair and maintenance, as well as contract supervision and inspection procedures. Such a PBC Action Program should be prepared for each CAREC member country, guiding future support to PBC development in the CAREC region.

www.ingramcontent.com/pod-product-compliance
Lightning Source LLC
Chambersburg PA
CBHW050050220326

41599CB00045B/7350